Commercial Aircraft Hydraulic Systems

The development of the book was sponsored by
Shanghai Jiaotong University Press

Commercial Aircraft Hydraulic Systems

Shanghai Jiao Tong University Press
Aerospace Series

Shaoping Wang
Department of Mechatronic Engineering
Beihang University, China

Mileta Tomovic
Batten College of Engineering and Technology
Old Dominion University, USA

Hong Liu
AVIC
The first Aircraft Institute

AMSTERDAM • BOSTON • HEIDELBERG • LONDON
NEW YORK • OXFORD • PARIS • SAN DIEGO
SAN FRANCISCO • SINGAPORE • SYDNEY • TOKYO

Academic Press is an imprint of Elsevier

ELSEVIER

Academic Press is an imprint of Elsevier
225 Wyman Street, Waltham, MA 02451, USA
The Boulevard, Langford Lane, Kidlington, Oxford OX5 1GB, UK

Notices
Knowledge and best practice in this field are constantly changing. As new research and
experience broaden our understanding, changes in research methods, professional practices,
or medical treatment may become necessary.

Practitioners and researchers must always rely on their own experience and knowledge in
evaluating and using any information, methods, compounds, or experiments described
herein. In using such information or methods they should be mindful of their own safety and
the safety of others, including parties for whom they have a professional responsibility.

To the fullest extent of the law, neither the Publisher nor the authors, contributors, or editors,
assume any liability for any injury and/or damage to persons or property as a matter of
products liability, negligence or otherwise, or from any use or operation of any methods,
products, instructions, or ideas contained in the material herein.

ISBN: 978-0-12-419972-9

British Library Cataloguing-in-Publication Data
A catalogue record for this book is available from the British Library

Library of Congress Cataloging-in-Publication Data
A catalog record for this book is available from the Library of Congress

For information on all Academic Press publications
visit our website at http://store.elsevier.com/

Working together
to grow libraries in
developing countries

www.elsevier.com • www.bookaid.org

Typeset by TNQ Books and Journals
www.tnq.co.in

Printed and bound in the United States of America

Contents

4. New Technology of Aircraft Hydraulic System

Foreword

In general, the flight control system is the critical system of an aircraft. The aircraft hydraulic actuation system and its power supply system are very important, related systems that directly influence aircraft flight performance and flight safety. Over the past several decades, aircraft system design focused predominantly on the design principle itself without considering the related system effects. The hydraulic power supply system provides high-pressure fluid to the actuation system; therefore, its characteristics and performance could influence the actuation system performance. On the other hand, the actuation system utilizes hydraulic power to drive the surfaces, the performance of which not only depends on the displacement control strategy but also on the power supply performance. This book focuses on the aircraft flight control system, including the interface between the hydraulic power supply system and actuation system, and it provides the corresponding design principle and presents the latest research advances used in aircraft design.

The aircraft hydraulic system evolved with the flight control system. Early flight control systems were purely mechanical systems in which the pilot controlled the aircraft surfaces through mechanical lines and movable hinge mechanisms. With the increase in aircraft velocity, the hinge moments and required actuation forces increased significantly to the point at which pilots had difficulty manipulating control surfaces. The hydraulic booster appeared to give extra power to drive the surfaces. With the increasing expansion of flight range and duration of flight, it became necessary to develop and implement an automatic control system to improve the flight performance and avoid pilot fatigue. Then, the electrically signaled (also known as fly-by-wire (FBW)), hydraulic powered actuator emerged to drive the aircraft control surfaces. Introduction of the FBW system greatly improved aircraft flight performance. However, the use of many electrical devices along with the flutter influence of the hydraulic servo actuation system led to a reliability problem. This resulted in wide implementation of redundancy technology to ensure high reliability of the FBW system. Increasing the number of redundant channels will potentially increase degree of fault. To achieve high reliability and maintainability, a monitoring and fault diagnosis device is integrated in the redundant hydraulic power supply system and redundant actuation system.

Modern aircraft design strives to increase the fuel economy and reduction in environmental impacts; therefore, the high-pressure hydraulic power supply system, variable-pressure hydraulic system, and increasingly electrical system are emerging to achieve the requirements of green flight.

This book consists of four chapters. Chapter 1 presents an overview of the development of the hydraulic system for flight control along with the interface between the flight control system and the hydraulic system. The chapter also introduces different types of actuation systems and provides the requirements of the flight control system for specification and design of the required hydraulic system. Chapter 2 introduces the basic structure of aircraft hydraulic power supply systems, provides the design principle of the main hydraulic components, and provides some typical hydraulic system constructions in current commercial aircraft. Chapter 3 introduces the reliability design method of electrical and mechanical components in the hydraulic system. The chapter provides comprehensive reliability evaluation based on reliability, maintainability, and testability and gives the reliability evaluation of the aircraft hydraulic power supply and actuation system. Chapter 4 introduces new technologies used in modern aircraft, including the high-pressure hydraulic power supply system, variable-pressure hydraulic power supply system, and new types of hydraulic actuators.

We thank all of the committee members of a large aircraft flight control series editorial board and all of the editors of Shanghai Jiaotong Press for their help and assistance in successfully completing this book. The authors are also grateful to Ms Hong Liu, Mr Zhenshui Li, and Mr Yisong Tian, who reviewed the book outline and contributed to the writing of this book. We are indebted to their comments. We should also mention that some of the general theory and structure composition were drawn from related references in this book; therefore, we would like to express our gratitude to their authors for providing outstanding contributions in the related fields. Finally, we hope that the readers will find the material presented in this book to be beneficial to their work.

Shaoping Wang
Mileta Tomovic
Hong Liu
July 2015

Preface

Aircraft design covers various disciplines, domains, and applications. Different viewpoints have different related knowledge. The aircraft flight control series focus on the fields that are related to the aircraft flight control system and provide the design principle, corresponding technology, and some professional techniques.

Commercial Aircraft Hydraulic Systems aims to provide the practical knowledge of aircraft requirements for the hydraulic power supply system and hydraulic actuation system; give the typical system structure and design principle; introduce some technology that can guarantee the system reliability, maintainability, and safety; and discuss technologies used in current aircraft. The intention is to provide a source of relevant information that will be of interest and benefit to all of those people working in this area.

Chapter 1

Requirements for the Hydraulic System of a Flight Control System

1.1 THE DEVELOPMENT OF THE HYDRAULIC SYSTEM RELATED TO THE FLIGHT CONTROL SYSTEM [1]

The flight control system (FCS) is a mechanical/electrical system that transmits the control signal and drives the surface to realize the scheduled flight according to the pilot's command. FCSs include components required to transmit flight control commands from the pilot or other sources to the appropriate actuators, generating forces and torques. Flight control needs to realize the control of aircraft flight path, altitude, airspeed, aerodynamic configuration, ride, and structural modes. Because the performance of the FCS directly influences aircraft performance and reliability, it can be considered as one of the most important systems in an aircraft.

A conventional fixed-wing aircraft control system, shown in Figure 1.1, consists of cockpit controls, connecting linkages, control surfaces, and the necessary operating mechanisms to control an aircraft's movement. The cockpit controls include the control column and rudder pedal. The connecting linkage includes a push–pull control rod system and cable/pulley system. Flight control surfaces include the elevators, ailerons, and rudder. Flight control includes the longitudinal, lateral-directional, lift, drag, and variable geometry control system.

Since the first heavier-than-air aircraft was born, it is the pilot who drives the corresponding surfaces through the mechanical system to control the aircraft, which is called the manual flight control system (MFCS) without

Commercial Aircraft Hydraulic Systems. http://dx.doi.org/10.1016/B978-0-12-419972-9.00001-2
1

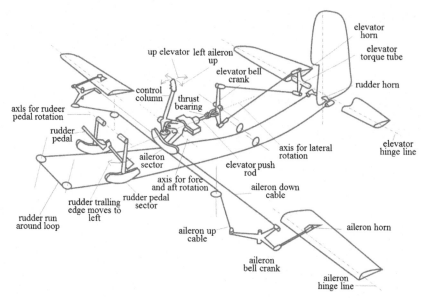

FIGURE 1.1 Structure of the initial FCS.

power. A very early aircraft used a system of wing warping in which no conventionally hinged control surfaces were used on the wing. A MFCS uses a collection of mechanical parts such as pushrods, tension cables, pulleys, counterweights, and sometimes chains to directly transmit the forces applied at the cockpit controls to the control surfaces. Figure 1.1 shows the aircraft's purely mechanical manipulating system, in which a steel cable or rod is used to drive the surfaces. If the pilot wants to move the flaps on a plane, then he would pull the control column, which would physically pull the flaps in the direction that the pilot desired. In this period, the designer focuses on the friction, clearance, and elastic deformation of the transmission system so as to achieve good performance.

With the increase of size, weight, and flight speed of aircraft, it became increasingly difficult for a pilot to move the control surfaces against the aerodynamic forces. The aircraft designers recognized that the additional power sources are necessary to assist the pilot in controlling the aircraft. The hydraulic booster, shown in Figure 1.2(a), appeared at the end of the 1940s, dividing the control surface forces between the pilot and the boosting mechanism. The hydraulic booster utilizes the hydraulic power with high pressure to drive the aircraft surfaces according to the pilot's command. As an auxiliary component, the hydraulic booster can increase the force exerted on the aircraft surface instead of the pilot directly changing the rotary or flaps. As the earliest hydraulic component that is

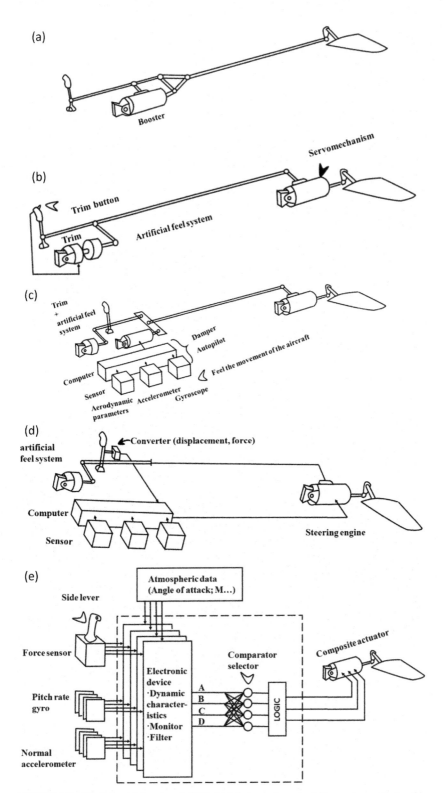

FIGURE 1.2 Evolution of the aircraft FCS. (a) Mechanical manipulating system with booster, (b) irreversible booster control system, (c) reversible booster control system, (d) stability augmentation control system, and (e) FBW systems [2].

related to the aircraft FCS, the hydraulic booster changed the surface maneuver from mechanical power to hydraulic power and resisted the hinge moment of surfaces without the direct connection between the control rod and surfaces. There are two kinds of hydraulic booster: reversible booster and irreversible booster. In the case of the irreversible booster control system shown in Figure 1.2(b), there is no direct connection between the control rod and the surface. The pilot controls the hydraulic booster to change the control surface without feeling of the flight state. The advantages of hydraulically powered control surfaces are that (aerodynamic load on the control surfaces) drag is reduced and control surface effectiveness is increased. Therefore, the reversible booster control system emerged through installing the sensing device to provide the artificial force feeling to the pilot, shown in Figure 1.2(c). The reversible booster control system includes the spring, damper, and additional weight to provide the feedback (feeling) so that a pilot could not pull too hard or too suddenly and damage the aircraft. In this kind of aircraft, the characteristics of booster (maximum output force, distance, and velocity) should satisfy the flight control performance.

In general, the center of gravity is designed forward of center of lift for positive stability. Modern fly-by-wire (FBW) aircraft is designed with a relaxed stability design principle. This kind of design requires smaller surfaces and forces, low trim loads, reduced aerodynamic airframe stability, and more control loop augmentation. This kind of aircraft operates with augmentation under subsonic speed. When the aircraft operates at supersonic speed, the aircraft focus moves backward, and the longitudinal static stability torque rapidly increases. At this time, it needs enough manipulating torque to meet the requirements of aircraft maneuverability. However, the supersonic area in the tail blocks the disturbance propagation forward, and the elevator control effectiveness is greatly reduced. Hence, it is necessary to add signals from stability augmentation systems and the autopilot to the basic manual control circuit. As we know, a good aircraft should have good stability and good maneuverability. The unstable aircraft is not easy to control. Because the supersonic aircraft's flight envelope expands, its aerodynamics are difficult to meet the requirements at low-altitude/low-speed and high-altitude/high-speed. In the high-altitude supersonic flight, the aircraft longitudinal static stability dramatically increases whereas its inherent damping reduces, then the short periodic oscillation in the longitudinal and transverse direction appear that greatly influences the aircraft maneuverability. To maintain stability of the supersonic aircraft, it is necessary to install the stability augmentation system shown in Figure 1.2(d). Because the stability augmentation system can keep the aircraft stable even in static instability design, the automatic flight control system (AFCS) appeared. The AFCS consists of electrical, mechanical, and hydraulic components that generate and transmit automatic control commands to the aircraft surfaces. Through measuring the perturbation from the gyroscope and accelerometer, the stability augmentation system generates the artificial damping with the help of reverse surface motion to quickly reduce the oscillation. The stability augmentation system provides good stability to the aircraft at high altitudes, high speeds, and at a large angle of attack states. In this kind of system, the stability augmentation is independent of the pilot manipulating system. To safely manipulate the aircraft, the stability augmentation and pilot manipulating system have different control limits of authority. From the pilot's point of view, the stability augmentation system is the part of aircraft and the pilot controls the aircraft like an "equivalent aircraft" with good control performance. Because the aircraft surface is controlled both by control column command and by augmentation system command, the control authority of augmentation system is just 3–6% of control authority.

Although the stability augmentation system can improve aircraft stability, it can also weaken the aircraft control response sensitivity to a certain extent, which will reduce its maneuverability. To eliminate this drawback, the control stability augmentation system emerges with the pilot's command based on the stability augmentation system shown in Figure 1.2(d). Through adjustment of the pilot control and control stability augmentation, the contradiction between stability and controllability can be solved to achieve good aircraft maneuverability and flexibility. Because the pilot can directly control the surface, the authority of augmentation can be increased to more than 30% of control authority.

In this period, the hydraulic actuators were used to drive the surfaces, which are powered by hydraulic pumps in the hydraulic circuit. The hydraulic circuit consists of hydraulic pumps, reservoirs, filters, pipes, and actuators. Hydraulic actuators convert hydraulic pressure into control surface movements.

Although the hydromechanical control system can realize the control with good stability and good maneuverability, it is difficult to realize fine manipulation signal transmission because of the inherent friction, clearance, and elastic deformation existing in the mechanical system. The following are common disadvantages for traditional mechanical systems or systems with augmentation:

1. The mechanical transmission and control system is big and heavy.
2. It has inherent nonlinear factors such as friction, clearance, and natural vibration due to hysteresis.
3. The mechanical control system is fixed in the aircraft body, which can lead to elastic vibration and could cause the control rod offset and sometimes vibration of the pilot

Then, in the early 1970s, FBW (Figure 1.2(e)) appears to overcome the above shortcomings. FBW cancels the conventional mechanical system and adopts an electrical signal to transmit the pilot's command to the control augmentation system. In brief, FBW is all full authority "electrical signal - plus control augmentation system" FCS, which transmits the pilot's command with electrical cable and utilizes the control augmentation system to drive the surface motion. In FBW, hydraulic actuation is the main component connected between flight controller and aircraft surfaces.

There are many advantages of FBW, including performance improvement, insensitivity to the aircraft structure unstable unfluence, and ease of connection with the autopilot system. However, this system was built to very stringent dependability requirements in terms of safety and availability. The following factors need to be considered when designing a FBW system.

1.1.1 Mission Reliability [3]

Mission reliability is defined as the probability of the system for being free of failure for the period of time required to complete a mission. The probability is a point on the reliability function corresponding to the mission length. The mission reliability of a system can be described as

$$R_M(t) = P(T > t_M) \tag{1.1}$$

where $R_M(t)$ is the mission reliability of system, P is the probability, T is the life of system, and t_M is the mission time.

In general, the reliability of FBW is not very high compared with the aircraft mechanical control system. Therefore, the reliability should be guaranteed when the FBW is used in aircraft. There are two indices to evaluate the aircraft reliability: flight safety and mission reliability. According to the aircraft control system design specification (MIL-F-9490D) [4], the probability of mission failure per flight due to relevant material failures in the FCS shall not exceed the applicable limit specified below [4].

1. Overall aircraft mission accomplishment reliability is specified by the procurement activity $Q_{M(FCS)} \leq (1 - R_M)A_{M(FCS)}$
2. Overall aircraft mission accomplishment reliability is not specified $Q_{M(FCS)} \leq 1 \times 10^{-3}$

Where $Q_{M(FCS)}$ is the maximum acceptable mission unreliability due to relevant FCS material failures, R_M is the specified overall aircraft mission accomplishment reliability, and $A_{M(FCS)}$ is the mission accomplishment allocation factor for flight control (chosen by the contractor).

Failures in power supplies or other subsystems that do not otherwise cause aircraft loss shall be considered where pertinent. A representative mission to which the requirement applied should be established and defined in the FCS

TABLE 1.1 FCS Quantitative Flight Safety Requirements

	Maximum aircraft loss rate from FCS failure
MIL-F-8785, class III aircraft	$Q_{S(FCS)} \leq 5 \times 10^{-7}$
All rotary wing aircraft	$Q_{S(FCS)} \leq 25 \times 10^{-7}$
MIL-F-8785 class I, II, and IV aircraft	$Q_{S(FCS)} \leq 100 \times 10^{-7}$

specification. If the overall aircraft flight safety in terms of R_S is not specified by the procuring activity, then the numerical requirements given in Table 1.1 apply [4].

1.1.2 Quantitative Flight Safety [4]

The probability of aircraft loss per flight due to relevant FCS material failures in the FCS shall not exceed $Q_{S(FCS)} \leq (1 - R_S)A_{S(FCS)}$ [4].

Where $Q_{S(FCS)}$ is the maximum acceptable aircraft loss rate due to relevant FCS material failures, R_S is the specified overall aircraft flight safety requirement as specified by the procuring activity, and $A_{S(FCS)}$ is the flight safety allocation factor for flight control (chosen by the contractor).

The maximum aircraft loss rate from FCS failures $Q_{S(FCS)}$ is as follows:

Class I and II aircraft: 62.5×10^{-7}/flight hour
Class III aircraft: 0.746×10^{-7}/flight hour

Likewise, the maximum aircraft task interruption rate from FCS failures $Q_{M(FCS)}$ is

Class I and II aircraft: 0.625×10^{-3}/flight hour
Class III aircraft: 0.15×10^{-3}/flight hour

At present, the safety requirement of an FCS is 1.0×10^{-7}/flight hour for military aircraft and $1 \times 10^{-9} \sim 1 \times 10^{-10}$/flight hour for commercial aircraft. To achieve such high reliability requirements, it is necessary to utilize the redundancy design method.

The overall reliability of the aircraft FBW system depends on the computer control/monitor architecture, which provides the tolerance to hardware and software failures, the servo control, and the power supply arrangement. Thus the redundancy, failure monitoring, and system protection emerged in the system design. The aircraft safety is demonstrated in the airworthiness regulation. In aircraft design, the faults, interaction faults, and external

TABLE 1.2 Flight Control Technology Chronology

Technology	Military	Commercial
Unpowered	1910s	1920s
Power boost	1940s	1940s
3000-psi hydraulics	1940s	1950s
Autopilots	1950s	1950s
Fully powered with reversion	1950s	1960s (Boeing 727)
Fully powered without reversion	1950s (B-47)	1970 (Boeing 747)
FBW	1970s (F-16)	1980s (A320)
Digital FBW	1970s	1980s (A320)
5000-psi hydraulics	1990s (V-22)	2005 (A380)
Power-by-wire	2006 (F-35)	2005 (A380)

environmental hazards should be considered. For physical faults, FAR/JAR 25.1309 provides the quantitative requirements.

Summarizing the above development of the aircraft FCS, its chronology can be seen in Table 1.2 [5].

1.2 THE INTERFACE BETWEEN THE FCS AND HYDRAULIC SYSTEM

Actuation systems are a vital link between the flight controls and hydraulic systems, providing the motive force necessary to move flight control surfaces. All of the flight controls need the force to drive the surface motion. Hydraulic actuators are the system that converts hydraulic pressure into control-surface movements. Because the performance of the actuation system significantly influences the overall aircraft performance, the aircraft will dictate some requirements in actuation system design.

1.2.1 Aircraft Control Surfaces [6]

The aircraft control system includes several different flying control surfaces, Figure 1.3, including primary control surfaces and secondary control surfaces. The primary flight control consists of elevators, rudders, and ailerons, which generate the torque to realize the pitch, roll, and yaw movements of the aircraft. The secondary flight control is in charge of the aerodynamic configuration of the aircraft through the control of the position of flap, slats, spoilers, and the trimmable horizontal stabilizer.

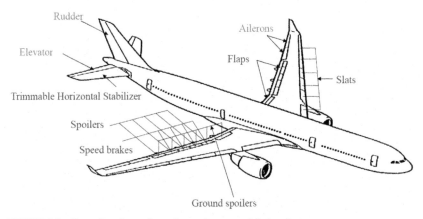

FIGURE 1.3   Control surfaces of an advanced commercial aircraft.

1.2.1.1 Primary Flight Controls [7]

A conventional primary control consists of cockpit controls, computers, connecting mechanical and electric devices, number of aerodynamic movable surfaces, and the required power sources. Primary flight controls include the pitch control, roll control, and yaw control shown in Figure 1.4. Primary flight

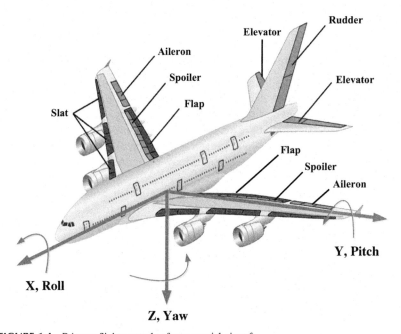

FIGURE 1.4   Primary flight controls of commercial aircraft.

control is critical to safety, and loss of control in one or more primary flight control axis is hazardous to the aircraft.

Pitch control is exercised by four elevators located on the trailing edge of the aircraft. Each elevator section is independently powered by a dedicated flight control actuator, which in turn is powered by one of several aircraft hydraulic power systems. This arrangement is dictated by the high integrity requirements placed upon FCSs. The entire tail section of the plane is powered by two or more actuators to trim the aircraft in pitch. In the case of emergency, this facility could be used to control the aircraft, but the rates of movement and associated authority are insufficient for normal control purposes.

Roll control is provided by two aileron sections located on the outboard third of the trailing edge of each wing. Each aileron section is controlled by a dedicated actuator powered by one of the aircraft hydraulic systems. At low airspeeds, the roll control provided by the ailerons is augmented by differential use of the wing spoilers mounted on the upper surface of the wing. During a right turn, the spoilers on the inside wing of the turn (i.e., the right wing) will be extended. This reduces the lift of the right wing, causing it to drop, thereby enhancing the desired roll demand.

Yaw control is provided by three independent rudder sections located on the trailing edge of the fin (or vertical stabilizer). These sections are powered in a similar fashion as elevators and ailerons. On a commercial aircraft, these controls are associated with the aircraft yaw dampers. They damp out unpleasant "Dutch roll" oscillations, which can occur during flight and that can be extremely uncomfortable for the passengers, particularly those seated at the rear of the aircraft.

1.2.1.2 Secondary Flight Controls [8]

Secondary flight controls include flap control, slate control, ground spoiler control, and trim control. Flap control is affected by several flap sections located on the inboard two-thirds of the wing trailing edge. Deployment of the flaps during takeoff or landing extends the flap sections rearward and downward to increase the wing area and camber, thereby greatly increasing lift for a given speed. The number of flap sections may vary among different types of aircraft.

Slat control is provided by several actuators, which extend forward and outward from the wing leading edge. In a similar fashion to the flaps described above, the slats have the same effect of increasing wing area and camber and therefore overall lift. A typical aircraft may have five slat sections per wing.

The ground spoiler serves as the speed-brake, which is deployed when all of the over-wing spoilers are extended together. The overall effect of the ground spoiler is reduced lift and increased drag. The effect is similar to the application of air-brakes in a fighter jet, where increasing drag allows the pilot to rapidly adjust aircraft airspeed; most airbrakes are located on the rear fuselage upper or lower sections and may have a pitch moment associated with their deployment. In most cases, compensation for this pitch moment would be automatically applied within the FCS.

1.2.2 Interface between Flight Controls and Hydraulic Systems

The development of the hydraulic system related to previously discussed flight controls indicates that the interface between flight controls and hydraulic systems is the actuation system shown in Figure 1.5, in which three hydraulic power supply systems (viz. green, yellow, and red) provide the power to the corresponding actuators. The performance of the actuation system directly affects the aircraft flying quality; therefore, the actuation systems play an important role in FCSs.

The interface between the hydraulic system and flight control is the hydraulic-powered actuator, which connects to control surfaces. Although different surfaces need a different number and type of actuator, the linkage between the hydraulic power supply and flight control is the actuator. Different flight control allocation has a different interface. In the case of the centralized hydraulic power supply system, the interface between the hydraulic power supply and flight control is the hydraulic actuator. Whereas in the case of the distributed flight controls, the interface between flight control and the hydraulic system is the electrohydrostatic actuator (EHA) [9] or the electrical mechanical actuator (EMA). To describe the relation with the hydraulic system, Figure 1.6 gives the interconnection diagram among different subsystems, in which the servo valve converts the pilot's electrical command to the large amount of

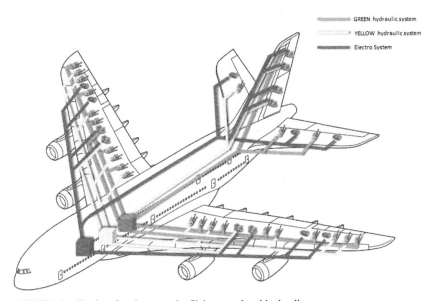

GREEN hydraulic system
YELLOW hydraulic system
Electro System

FIGURE 1.5   The interface between the flight control and hydraulic systems.

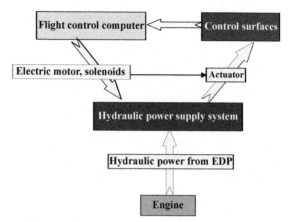

FIGURE 1.6 The relationship between the FCS and the hydraulic system [10].

power delivered to the actuators with the high-pressure hydraulic power delivered. So the interface between flight controls and hydraulic system is actuator powered by hydraulic power supply system [11,12].

Airbus FBW systems adopt the five full-authority digital computers controlling the pitch, yaw, and roll and a mechanical backup on the trimmable horizontal stabilizer and the rudder. Figure 1.7 shows the flight control surfaces of the A320 family, in which ELAC indicates the elevator aileron computer, SEC indicates the spoiler elevator computer, and FAC indicates the flight augmentation computer [6]. The FBW system depends on the hydraulic-powered actuators to move the control surfaces and on the computer system to transmit the pilot controls. The pressurized servo control actuator is powered by three hydraulic circuits (green, yellow, and blue), where each one is sufficient to control the aircraft. One of the three circuits can be pressurized by the ram air turbine (RAT), which can be switched on when all engines flame out. The electrical power is supplied by two separate networks, each driven by one or two generators. If the normal electrical generation fails, then an emergency generator supplies power to a limited number of flight control computers. The last of these flight control computers can also be powered by two batteries.

The actuation system is a key element in an FCS because it links the input signal/input power and transfers it to drive the control surfaces shown in Figure 1.8.

It is obvious that the interface between flight control and the hydraulic power supply system is hydraulic power actuation, in which the servo valve is the key element that can convert the electrical signal to hydraulic power. There are several types of actuation systems powered by centralized hydraulic supply, such as the simple mechanical/hydraulic actuator, the mechanical actuator with electrical signal, and multiple redundant hydraulic-powered actuators.

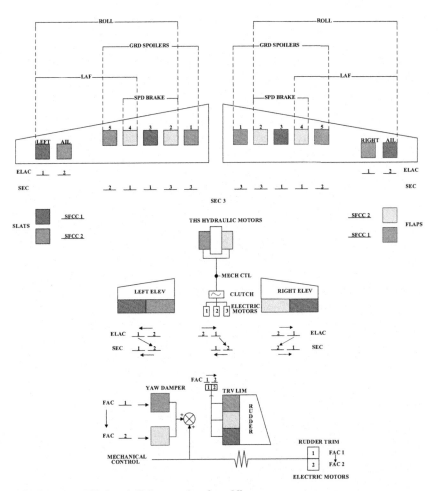

FIGURE 1.7   A320 aircraft flight control surfaces [6].

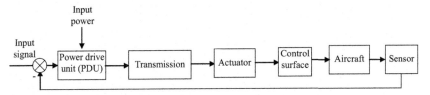

FIGURE 1.8   The structure of the actuation system.

1.3 ACTUATION SYSTEMS

The study of aircraft FCS development indicates that the interface between the FCS and hydraulic system is the actuation system. The actuation system plays an important role in attaining the specified performance of FCSs. There are several different types of actuation systems used in the current aircraft:

- Simple mechanical/electrical signaled, central hydraulic supply powered
- Multiple redundant electrohydraulic actuation
- Simple electrical signaled, distributed hydraulic supply powered

1.3.1 The Actuation System Powered by Centralized Hydraulic Supply [6,10,13]

Since the 1950s, the actuation system powered by centralized hydraulic supply was designed to maneuver the surface movement. Hydraulic fluids are used primarily to transmit and distribute forces to various units to be actuated. The early actuators were mechanical, Figure 1.9, in which the demand signal drives a spool valve and opens ports with high-pressure hydraulic fluid. The fluid enters the plunger cavity of a cylinder, pushes the piston rod to extend or retract, and drives the control-surface motion. When the spool valve moves to the required position, the mechanical feedback will close the valve and the cylinder movement stops. The hydraulic servo valve converts hydraulic power to drive the control surface through adjusting the nozzle opening. The aircraft response is feedback to the pilot.

Development of the FBW system allowed the actuator to utilize the electrical signals in conjunction with hydraulic power, Figure 1.10. Hydraulic actuators are widely used in commercial aircraft surface control because of their numerous advantages:

1. Fluids are almost incompressible
2. High-pressure fluid can deliver high forces
3. High power per unit weight and volume
4. Good mechanical stiffness
5. Fast dynamic response

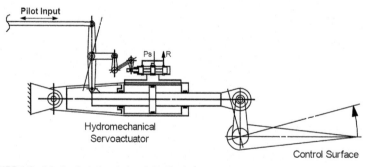

FIGURE 1.9   Mechanical signaled and feedback actuator.

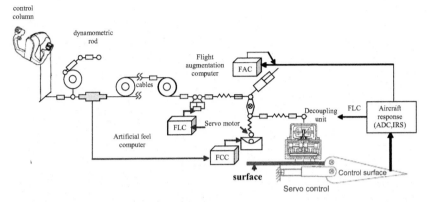

FIGURE 1.10   Mechanical flight control with actuator powered by centralized hydraulic supply.

The electrical command causes the hydraulic servo valve to open the spool shown in Figure 1.11. The high-pressure fluid enters the cylinder, moves the piston, and forces control surfaces to move to the desired position. In case of failure, the bypass valve allows the surface to be controlled freely by another actuator.

Because the electrohydraulic servo valve has a torque motor and hydraulic amplifier, its reliability is not very high. In most cases, the reliability of the hydraulic power supply system is higher than the electrical part, so the level of redundancy refers to the number of electrical parts used and not the number of hydraulic supplies. The common technology is to adopt redundancy in

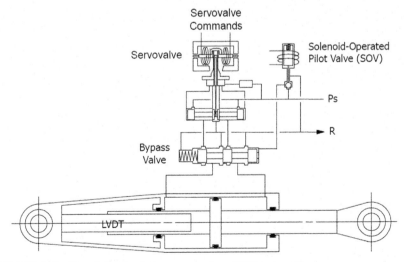

FIGURE 1.11   Electrically signaled servo valve in an FBW system [6].

electrical parts to greatly improve the system reliability. Therefore, the redundant actuator based on the number of servo valves or motor coils is widely used in aircraft FCSs. Figure 1.12 shows that the quad redundant electrical channels are designed with quad servo valve and shutoff valve coils and quad servo valve and linear variable differential transformers (LVDTs) [14]. The dual independent hydraulic supplies are integrated with an actuator ram of tandem construction. To maintain high reliability and safety, the two hydraulic power supply systems are designed separately and the actuator can accept the hydraulic power from each hydraulic power supply system. If one of the hydraulic-supply systems fails, the remaining hydraulic power supply system will continue to provide enough power to move the actuator against air loads. However, the movement of the ram will cause hydraulic fluid flow into and out of the cylinders on the side of the faulty hydraulic supply, which could create a drag force to prevent the ram movement. Bypass valves are designed in the actuator to connect the two sides of the cylinder in the event of loss of hydraulic pressure. A rip-stop ram design of the actuator is used to ensure that fatigue damage in one side of the cylinder will not cause a crack in the other side of the cylinder.

Figure 1.13 shows a redundant actuator with a tandem main control spool valve which is used to provide the motive force for the servo valve. This particular actuator uses four servo valves to drive the main spool valve, each

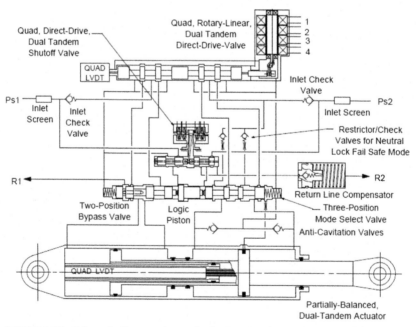

FIGURE 1.12   Schematic diagram of a quad redundant electrical channel actuator [6].

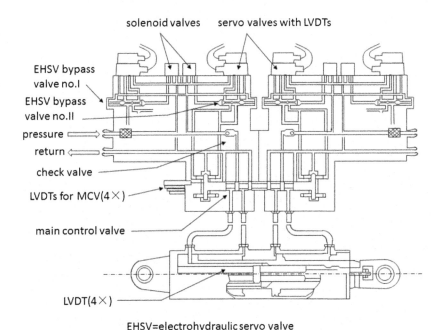

EHSV=electrohydraulic servo valve
LVDT=linear variable differential transformer
MCV=main control valve

FIGURE 1.13   Structure of a typical hydraulic servo actuator.

signaled by one of four flight control computers and four LVDTs which are used to measure main ram displacement. The high-pressure fluid enters the cylinder to produce the force of a quadruplex redundant actuator. The monitoring system compares each of the four signals to detect and isolate the failed lanes. If one or two lane fails at a time, then the monitoring system adopts a majority vote to meet system safety requirements.

The reliability of the actuation system is very important for flight controls; therefore, the redundancy techniques are necessary in primary actuators to ensure continued operation after a failure to meet the fail-operation-fail-operation requirement in actuator design. Modern aircraft primary flight controls have adopted quadruplex flight control computers and quadruplex actuators, in which feedback sensors are quadruplexed. The four flight control computers compare signals across a cross-channel data link to identify whether any of the signals differ significantly from the others. A consolidated or average signal is produced for use in control and monitoring algorithms, and each flight control computer (FCC) produces an actuator drive signal to one of the four coils in the direct drive valve motor, which moves the main control valve to control the tandem actuator [14].

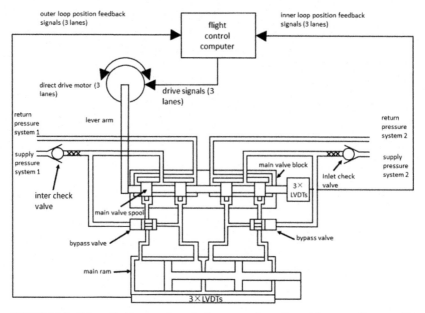

FIGURE 1.14   Schematic diagram of a typical actuator using a direct-drive-motor first stage [6].

Another type of actuator for which the first stage is driven directly by a motor is shown in Figure 1.14. The actuators use a rotary brushless DC motor to convert rotary motion to linear motion of the main control valve through a crank mechanism. This kind of actuator uses three coils in the direct drive motor and three feedback sensors (LVDTs) for each main control valve and main ram. The triplex actuator can operate even under the conditions of two similar but independent electrical failures. With the self-monitoring in lane, it can achieve fail-operation-fail-operation.

With the increase of aircraft velocity, the hinge moment of control surfaces changes greatly in the entire flight profile envelope. Thus, it is of no practical significance to use the reversible booster FCS. Especially after aircraft breaks through the sound barrier, the efficiency of the control surface sharply declines, and the focus of the aerodynamic load rapidly moves backward. To compensate for the overcompensation in subsonic conditions, the irreversible booster control system was developed. In this situation, the pilot cannot feel the hinge moment of control surface; therefore, it is difficult for the pilot to control the aircraft. The artificial feeling system appears to provide the control surface feeling. With the increasing expansion of flight range and duration of flight, it was necessary to provide an automatic control system to improve the flight performance and avoid pilot fatigue. As a result the electrical signaled hydraulic-powered actuator emerged to drive the control surfaces of aircraft.

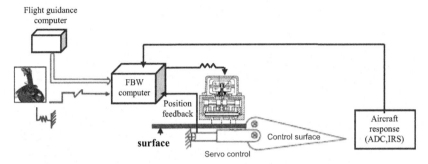

FIGURE 1.15   FBW control with the actuator powered by centralized hydraulic supply.

The electrical FCS, also called FBW, Figure 1.15, utilizes the electrical channel to replace complex mechanical transmission. The pilot's command and autopilot control signal are integrated in the computer that generates the driven signal sent to the servo control of the actuators at each aerodynamic surface. This solution was first designed in the 1960s and was utilized rapidly afterward. Moreover, the computer can also perform the necessary computation for augmentation function without the pilot's attention. In this case, the control signals to the aerodynamic surfaces are transmitted by electrical wiring.

Figure 1.16 is the FBW primary surface actuator schematic (with damped fail-safe mode). There are two modes in this kind of actuator:

- Active mode: actuator motion responds to the electrical command to the servo valve
- Damped mode: cylinder chambers are connected together through an orifice, the actuator moves with external force, damping suppresses flutter, and a compensator provides emergency fluid.

The principle of actuation system is described in the following subsections.

1.3.1.1 Mechanical/Hydraulic Actuator [15,16]

Figure 1.17 shows the conventional linear actuator powered by a dual hydraulic power supply system (viz. blue channel and green channel). In this type of actuator, the mechanical signal and the electrical signal can act on the summing link of the actuator, in which the servo valve (SV) converts the electrical command to the movement of the ram with the high-pressure hydraulic fluid supply. As the ram moves, the feedback link will rotate the summing link about the upper pivot, returning the servo valve input to the null position as the command position is achieved. The performance of the hydraulic actuator is to satisfy the demand with the hydraulic power-assisted mechanical response.

Because the hydraulic actuator is able to accept the hydraulic power from two identical/redundant hydraulic-supply systems, the aircraft control can maintain the function even in the case of loss of one fluid or a failure in any

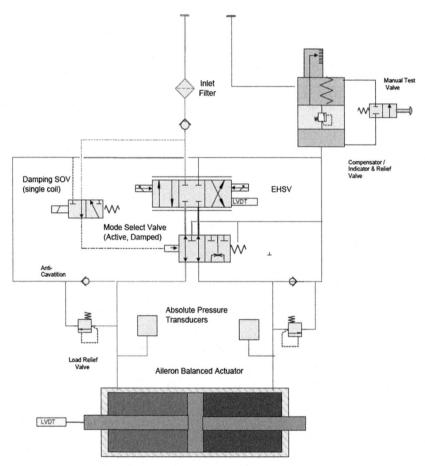

FIGURE 1.16   FBW flight control actuator with fail-safe mode.

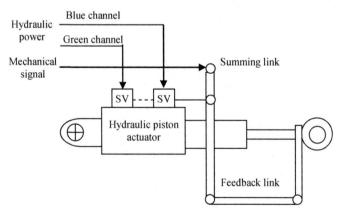

FIGURE 1.17   Conventional linear actuator [6].

one hydraulic power supply system. Likewise, the control surfaces can be driven even in the case of loss of one or more actuators; however, compensations should be built in to take control in the case of control surface actuator failure. In order to avoid affecting the aircraft flight performance, the actuators themselves have a simple reversion model-following failure; that is, return to the central position under the influence of aerodynamic forces. This reversion mode is called *aerodynamic centering* and is generally preferred over a control surface freezing or locking at some intermediate point in its travel. In some systems, "freezing" the FCS may be an acceptable solution depending upon the control authority and reversionary modes that the FCS possesses. The decision to implement either one of these approaches depends on the system safety analysis.

Mechanical actuation may also be used for spoilers in case the displacement of closed-loop control forces the spoiler surface to the closed position under the failure mode of aerodynamic closure. It will have no adverse effect upon aircraft handling.

1.3.1.2 Electrical Signaled Actuator

Although mechanical actuation has already been widely used in many applications, most of the modern aircraft have electrically controlled and hydraulically powered redundant actuators. The demands for electrohydraulic actuators fall into two categories—simple demand signal or auto-stabilization inputs, shown in Figure 1.18.

Aircraft autopilot is used to reduce pilot workload, and its command can be coupled by pilot's command to the actuator. Pilot input to the actuator acts as manual manipulation. In the case when the autopilot is engaged the electrical input takes precedence over the pilot's demand. The actuator operates in an identical fashion as before with the mechanical inputs to the summing link causing the servo valve to move. The pilot could retrieve control through

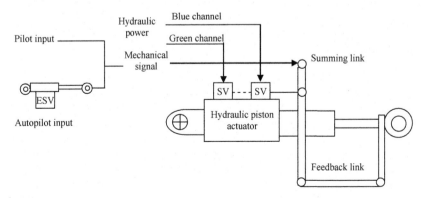

FIGURE 1.18 Conventional linear actuator with an autopilot interface [6].

disengaging the autopilot, and then the aircraft control system is restored to manual control.

Commonly, pilot gives a command in the form of electrical signal to the hydraulic servo valve, which then drives the hydraulic cylinder to move the control surface. However, for certain noncritical flight control surfaces, it may be faster less expensive and lighter to utilize an electrical link instead of hydraulic transmission system. In general, the electrical demand is stabilization signal derived within computer unit. The simplest form of autostabilization is the yaw damper, which damps out the cyclic cross-coupled oscillations that occur in roll and yaw known as "Dutch roll."

1.3.1.3 Multiple Redundancy Actuation System

Modern FCSs are increasingly adopting a FBW solution to reduce the weight and improve handling performance. Because the reliability of the electrical component is lower than the one of the mechanical component, multiple redundancy electric signaling with simplex hydraulic supply is incorporated in the FBW system, Figure 1.19 [6].

Multiple redundant electrohydraulic actuators are shown in Figure 1.20, [6] in which the slants indicate redundant electrical signals. Two slants express double lanes and four slants describe quadruplex identical lanes. When the solenoid valve is energized, it supplies hydraulic power to the actuator, often from two hydraulic systems. Control demands from the flight control computers are fed to the servo valves. The servo valves control the position of first-stage valves that are mechanically summed before applying demands to the control valves. The control valves modulate the position of the control ram. LVDTs measure the position of the first-stage actuator and output ram position of each lane, and these signals are fed back to the flight control computers. Hence, the servo valve serves as the interface between flight control and the hydraulic power supply.

Common characteristics of conventional electrohydraulic actuators include the following:

- Efficient in terms of dynamic response, but not efficient in terms of energy
- Can generate high force, low speed using direct drive

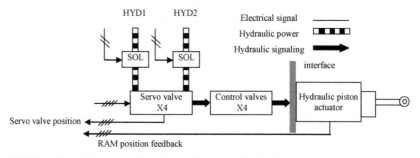

FIGURE 1.19   Multiple redundant electrically signaled hydraulic actuator.

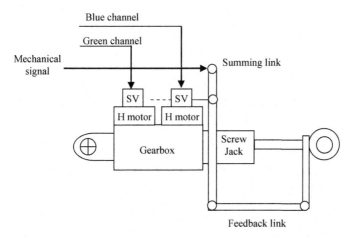

FIGURE 1.20   Mechanical screwjack actuator.

- Expensive maintenance, risk of pollution, risk of leakage, and risk of fire
- Easily integrated in the structure as jacks
- Easily designed for parallel redundancy
- Require constraining centralized hydraulic source at constant pressure filling, bleeding, and aging

  Power-by-wire actuators have the following characteristics:

- Easy power management
- Reduced installation constraints
- Must be designed with care with respect to electromagnetic interference (EMI) and temperature
- Produce fast-response/low-magnitude forces

1.3.1.4 Mechanical Screwjack Actuator [17]

The conventional actuator provides fast response for aileron, elevator, and rudder controls at light aerodynamic loads. If the aircraft requires slow response and large loads, then the mechanical screwjack actuator is required to withstand the aerodynamic loads, Figure 1.20, such as in the case of the trimmed horizontal stabilizer (THS). The THS provides the slow movement over small angular displacements for large loads in trimming the aircraft in pitch as airspeed varies.

The mechanical screwjack actuator has multiple hydraulic power supply systems. Two servo valves control the flow of fluid to the hydraulic motor, which in turn drives the screwjack by means of the gearbox. When the output of the screwjack actuator satisfies the pilot's demand, the movement of the ram leads the feedback to null the original demand.

1.3.2 The Actuation System Powered by Electrical Supply [6,18,19]

Although centralized hydraulic power actuators can provide high-frequency and good dynamic performance, they have a large topology with long pipes and centralized power supply systems. Rapid development of powerful new AC electrical systems paved the way for the actuation system powered by the aircraft AC electrical supply. Figure 1.21 is the schematical representation of an actuator powered by electrical supply. The actuator is powered by a constant-frequency, split-parallel, 115-VAC, three-phase electrical system. The three-phase constant speed electrical motor accepts the operational demand and drives the variable displacement hydraulic pump. The hydraulic pump provides the hydraulic pressure to power the actuator motion.

In Figure 1.23, the electrical motor works as a control unit and the hydraulic pump acts as a transmission unit. The variable displacement hydraulic pump provides the high-pressure source for the actuator. The servo valve can control the pump flow and actuator velocity through a bidirectional displacement mechanism. The displacement of the actuator is fed back to the servo valve to achieve the desired output position.

There are several types of actuators powered by electrical supply, as discussed in the following subsections.

1.3.2.1 Electrohydrostatic Actuator [6,20]

The problem with hydraulic systems is that they are heavy and they require significant space and much maintenance. An electrohydrostatic actuator (EHA) is used to resolve those issues by eliminating central pumps and hydraulic pipes. EHA is a new type of actuator that uses power electronics and control techniques to provide efficient flight control actuation. A conventional hydraulic actuator is continually pressurized by a centralized hydraulic power

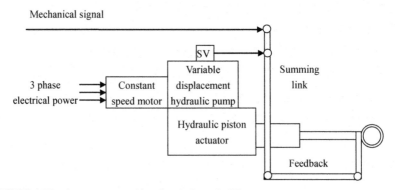

FIGURE 1.21   Actuator powered by electrical supply [6].

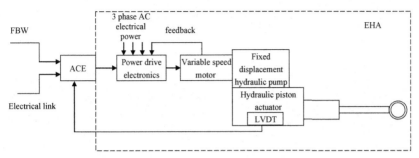

FIGURE 1.22   Electrohydrostatic actuator [6].

supply system whether or not there is any demand whereas the actuator demands are minimal in many cases. In this situation, most of the energy coming from the hydraulic power supply system is converted to heat through the orifice. Therefore, the constant pressure supply actuator wastes the energy from the engine and has a very high fuel consumption. The more-electrical aircraft provides another kind of electrical actuator, the EHA, which provides the more efficient actuation form according to the control demand. EHA uses three-phase AC power to feed power drive electronics, which in turn drives the constant-displacement hydraulic pump and makes the cylinder movement shown in Figure 1.22. The EHA uses the local hydraulic system, which reduces the need for long pipes between the centralized hydraulic power supply and the actuator and thus decreases the corresponding weight. In addition, in case of no demand, the only power requirement of EHA is to maintain the control electronics. When the actuator control equipment sends the command, the power rapidly acts on the electronics to drive the variable speed motor and pressurizes the actuator resulting in the corresponding surface movement. Once the output of the surface satisfies the demand, the power electronics resumes its normal dormant state. The ACE electrically closes the control loop around the actuator.

An EHA can drive the surfaces according to demand; therefore, it is widely used in modern aircraft such as an Airbus A380 and Lockheed Martin F-35. In those aircraft, the matrix converter can convert the three-phase AC power from the 115-VAC electrical system to the 270-VDC system. EHA utilizes the 270 VDC to drive the brushless DC motor, which in turn drives the fixed displacement pump and results in cylinder movement. The advantage of EHA lends to a greater use of electrical power, and more-electrical aircraft/all-electrical aircraft becomes a reality.

1.3.2.2 Electromechanical Actuator [6]

The EMA, Figure 1.23, is another type of actuator that uses an electric motor and gearbox assembly to drive the surface movement instead of the electrical signaled and hydraulic-powered actuation system.

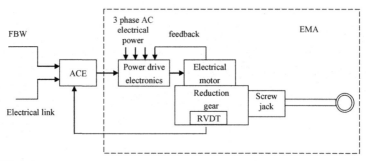

FIGURE 1.23   The structure of electro-mechanical actuator (EMA).

TABLE 1.3 Typical Applications of Different Actuators [6]

Actuator type	Power source	Primary flight control	Spoilers	THS	Flaps and slats
Linear actuator	Hydraulic system B/Y/G or L/C/R	X	X		
Screwjack actuator	Hydraulic or electrical system			X	X
Electrical signaled hydraulic actuator	Electrical systems and hydraulic systems	X	X		
EHA	Electrical system	X	X		
EMA	Electrical system			X	X

Note: B, Blue; Y, Yellow; G, Green; L, Left; C, Center; R, Right.

EMA uses the power drive electronics to drive the brushless DC motor, and the DC motor acts on the reduction gear to drive the surface movement. The structure of EMA is simple compared with the electrical signaled and hydraulic-powered actuator. However, its motive force is smaller and the response time is longer than the required flight controls; therefore, it is used in trim and door actuation. The disadvantage of EMA is the possibility of the actuator jamming, which restricts its application in the primary controls of aircraft.

Table 1.3 shows the applications of the different actuators.

1.3.3 Commercial Aircraft Implementation

The FBW system is widely used in commercial aircrafts. Concorde was the first commercial aircraft with the FBW system, and Airbus 320 was the first aircraft in the Airbus family that used the FBW system. Since then, the Airbus family and Boeing family have adopted FBW in modern commercial aircrafts. There are several differences in actuation system design philosophies, as presented in the following subsections.

1.3.3.1 The Flight Controls—Level Comparison [6,21,22]

High reliability and safety of the commercial aircraft are achieved through application of the redundant structure and monitoring system in flight controls. Boeing design philosophy adopts the similar lanes of three primary flight computers (PFCs) with dissimilar hardware and the same software, Figure 1.24(a). Each lane operates separately, and the voting system is used to detect discrepancies and disagreements among the lanes. The applied comparison decision logic varies for different types of data; for example, the median selector is used under quadrex lanes. The redundant ACE communicate with each other and directly drive the flight control actuator.

The Airbus philosophy adopts dissimilar redundant techniques, Figure 1.24(b), in which five separate main computers are used: three primary flight control computers and two secondary flight control computers. Each computer adopts different hardware and software. The command outputs from the secondary flight control computers are just for the standby use of ailerons, elevators, and rudders.

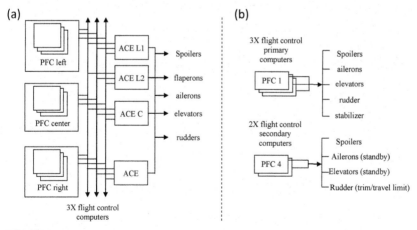

FIGURE 1.24 Boeing design philosophy versus Airbus design philosophy in FCS: (a) Boeing 777 top-level structure and (b) Airbus top-level structure [6].

1.3.3.2 Airbus Architecture [6]

The control surfaces of A320 are all hydraulic powered, Figure 1.25, in which the FBW system adopted a 7X2 flight control computer architecture. The system consists of two elevator aileron computers (ELACs), three spoiler elevator computers (SECs), two flight augmentation computers (FACs), two flight control data computers (FCDCs), two flight management guidance computers, and two slat flap control computers.

ELAC_2 provides the control of elevator, aileron, and the two electro-motors of the THS under normal conditions. In the case of ELAC_2 failure, ELAC_1 automatically substitutes for ELAC_2. In the case when both ELAC_1 and ELAC_2 fail, SEC_1 or SEC_2 would take over the control. The SEC controls all of the spoiler and standby elevator actuators (third THS). The FAC provides the conventional yaw damper function with the yaw damper actuators and realizes the automatic trim limit value monitoring function. FCDC sends data from the ELAC and SEC to the electrical instrument system and central failure display system. In addition, THS can provide safe landing even under all electrical system failure through the mechanical controls.

The surface allocation is as follows:

- Pitch controls:
  - Electrical control elevators
  - 1 mechanical control THS
- Roll controls:
  - Electrical control ailerons
  - Electrical control spoilers No.2—No.5

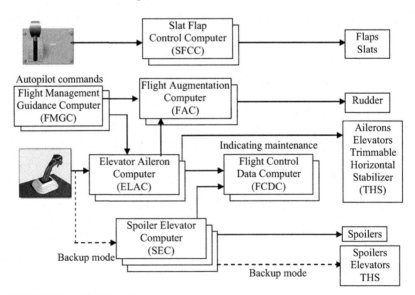

FIGURE 1.25   A320 FCSs [6].

- Yaw controls:
  - 1 mechanical control rudder
  - 1 yaw damper
- Speed breaks:
  - Electrical control spoilers No.2—4
- Ground spoilers:
  - All electrical control spoilers
- High lift
  - 10 electrical control slats
  - 4 electrical control flaps
  - 10 lift dumpers

There are three independent hydraulic power supply systems in A320 aircraft—blue (B), green (G), and yellow (Y). All the three systems provide high-pressure hydraulic power to the flight control actuators. The fundamental design principle is that the aircraft must fly and land even in the very unlikely event of failure of all computers. In this condition, the THS and rudder can be controlled directly to realize the pitch and lateral control of the aircraft by the mechanical trim input.

The A330/340 FBW system inherited the design principle of A320. The FCS consists of three flight control primary computers and two flight control secondary computers without a special FAC, Figure 1.26 [6]. The pitch control and rudder control still retain mechanical manipulation.

Every flight control computer has two channels: a command channel and a monitoring channel. The command channel operates the allocation mission whereas the monitoring channel guarantees the command correction. Flight control computers adopted dissimilar redundant microprocessors and totally different control software. The flight control primary computer (FCPC) utilizes an Intel 80386 microprocessor, in which Assembly language is adopted for the command channel, whereas PL/M language is adopted for the

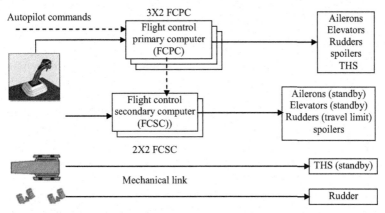

FIGURE 1.26   The structure of A340 flight control system (FCS).

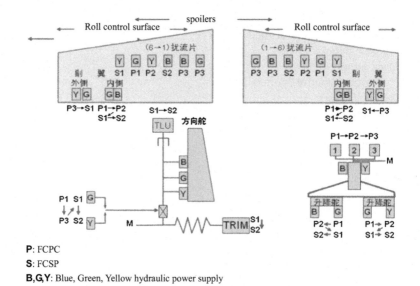

P: FCPC

S: FCSP

B,G,Y: Blue, Green, Yellow hydraulic power supply

FIGURE 1.27   A340 control surfaces allocation and control permissions change order [6].

monitoring channel. An Intel 80186 microprocessor is used for a flight control secondary computer (FCSC), where Assembly language is adopted for the command channel and the Pascal language is adopted for the monitoring channel. A340 aircraft control surface allocation is shown in Figure 1.27, where three hydraulic power supply systems (blue (B), green (G), and yellow (Y)) provide the hydraulic power to the flight control actuators.

The A380 FCS has adopted a new double architecture system, namely 2H+2E, in which the main control command is transmitted with the FBW system and power is transmitted with the power-by-wire system. This particular design philosophy, Figure 1.28, has been used in Airbus over the past 20 years, in which the centralized hydraulic power supply system and distributed electrical power supply system are simultaneously used. For example, the A380 aircraft aileron control adopts conventional hydraulic actuator (HA) power by the centralized hydraulic power supply and the EHA is powered by the electrical supply system at the same time. Under normal conditions, the HA actively drives the aileron and the EHA follows the HA. When the HA fails, the EHA drives the aileron instead of the HA.

A380 aircraft have moved away from mechanical control by replacing it with an electrical backup hydraulic actuator (EBHA). Although there is no mechanical control channel in A380, its completely dissimilar redundant design makes it reliable and safe. Application of EHA eliminates pipes between the centralized hydraulic power supply and the actuator, thus reducing the weight and improving the safety, efficiency, reliability, and maintainability of the aircraft. The electrical powered flight-by-wire actuator is becoming

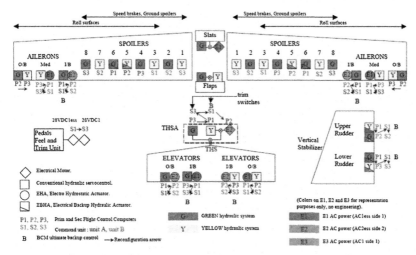

FIGURE 1.28   A380 structure layout based on 2H+2E [6].

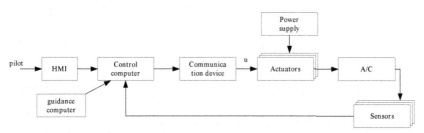

FIGURE 1.29   General architecture scheme of flight control channel.

increasingly used for primary flight controls. As its reliability has improved, the power-by-wire actuation systems has eliminated hydraulic systems, shown in Figure 1.29.

The bigger the aircraft, the more the actuators are adopted in order to maintain high reliability. Table 1.4 lists the number of spoilers and aileron actuators in the Airbus family.

With the progression of the Airbus family from A320 to A380, one can easily identify the integration trends. In A320 aircraft, the autopilot and flight management system are designed separately. In the case of the A330/340, the flight management and guidance computer combines the autopilot and guidance. A380 integration has progressed through synthesis of different kinds of flight control functions with stand-alone flight control computer.

In A380 aircraft, the flight control actuator configuration is quite different from that of A320 and A340. In A380 aircraft, both the electrical signaled HA and the EHA are used at the same time. Some actuators are powered by the central hydraulic power supply system (viz. green hydraulic system powered

TABLE 1.4 Airbus Family Actuator Number [6]

Airbus model	Spoilers per wing	Ailerons/actuators per wing
A320 family	5	1/2
A330/A340 family	6	2/4
A380	8	3/6

by engines 1 and 2 and yellow hydraulic system powered by engines 3 and 4). However, many are powered by the combination of HAs and EHAs. The following are the list of various actuation types used in A380 aircraft:

- Two outboard aileron surfaces and six spoiler surfaces on each wing are powered by HA.
- The inboard aileron surfaces and elevator surfaces are powered by both HA (primary) and EHA (backup).
- The middle two spoiler surfaces (5 and 6 on each wing) and rudder are powered by the EBHA.
- The THS actuator is powered by the hydraulic power supply systems (yellow and green) and electrical power supply system E2.
  There are two modes of the A380 aileron control:
  - Normal-HA mode: In the normal condition, the actuator receives hydraulic power from green or yellow systems and the servo valve drives the actuator according to the FBW computer demand.
  - Backup-EHA mode: In the case of HA failure, the actuator receives electrical power from the AC electrical system and controls motor rotation. The motor drives hydraulic pump rotation to pressurize the cylinder. The cylinder moves with high-pressure fluid to control the aileron.

1.3.3.3 Boeing 777 Implementation [6]

The design approach of the Boeing family of aircraft is to adopt a similar redundant FCS. Boeing started to partially use FBW in Boeing 777, in which 3X3 PFCs and actuator control electronics (ACE) are used with 629 bus aircraft, Figure 1.30. Various combinations of dissimilar hardware, different component manufactures, dissimilar control/monitoring functions, different hardware/software design teams, and different compilers are implemented in aircraft design.

The primary FCS of Boeing 777 aircraft consists of the following control-surface actuators:

- 4 elevators: left and right inboard and outboard
- 2 rudders: upper and lower

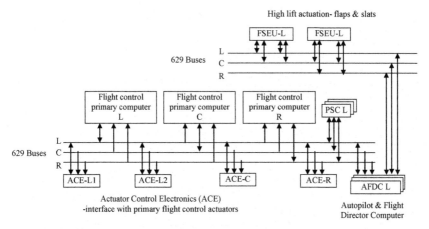

FIGURE 1.30   B777 primary flight control systems (FCSs) [6].

- 4 ailerons: left and right inboard and outboard
- 4 flaps: left and right inboard and outboard
- 14 spoilers: 7 left and 7 right

The flight control actuators are connected to the triple 629 flight control data buses through four actuator control electronics which contain digital-to-analog and analog-to-digital units. The FCS receives commands from the pilot or autopilot, and it controls the elevators, rudders, ailerons, flaps, slats, and HTS movement. The actuator ram position is fed back to the ACE to realize the aircraft controls. The pilot, through a mechanical link, can control the outer two spoilers to realize the roll control, and HTS to realize the pitch control. This function can guarantee a safe landing even in the case when all of the electric systems fail. In the case of Boeing 777, all actuators are powered by centralized hydraulic power supply systems.

1.4 REQUIREMENT OF THE FCS TO THE HYDRAULIC SYSTEM [23]

Actuation systems are a vital link between the flight controls and hydraulic systems, providing the motive force necessary to move flight control surfaces [7]. Whether it is primary flight controls or secondary flight controls, it is necessary to provide some force for moving the surface. The actuator performance directly influences the performance of the aircraft. The design of an aircraft FCS at all operating conditions must first consider the implications of actuator performance. According to the specified requirement of aircraft, arrangement and interconnection of hydraulic power sources and consumers is organized in a manner that meets requirements for controllability of aircraft. There are numerous requirements that need to be considered when designing the hydraulic system for flight control, some of which are discussed in the following subsections.

1.4.1 System Safety Requirements [4,24–27]

Primary flight control is critical for the safety of an aircraft. Loss of any one or more of the primary flight controls is hazardous for the aircraft. The active control technique used to maintain aircraft static stability in relaxed stability aircraft results in even greater reliance on the availability of primary flight control surfaces. Considering the safety requirement of the FCS, the actuation systems should at least be designed with a failure-operation-failure operation-failure-safe philosophy. In other words, the actuation system should operate at, or very close to, full performance when one or two failures appear to meet the safety and integrity requirements.

For the secondary control surfaces, the safety requirement will be somewhat lower. In this instance it is not necessary to ensure full operation under failures. In general, loss of operation of the secondary surface will not directly lead to the loss of aircraft. Therefore, the safety requirement of secondary control-surface always adopts a failure-operation-failure-safe philosophy. Of course, the actuator should be moved to the centralized position after the failures.

Most flight control actuation systems on current aircraft are electrically signaled and hydraulically powered. The demand drives the spool valve and opens the ports, and then the high-pressure hydraulic fluid flows in the cavity of the cylinder to extend or retract the control surface's movement. When the actuator approaches the required position, the mechanical feedback is used to close the valve. To realize the above safety requirement, the hydraulic power supply systems, actuators, and feedback LVDT should be designed with redundant techniques.

Hydraulic system design philosophy is as follows [10]:

- Multiple independent centralized hydraulic power supply systems: There should be more than one hydraulic system to ensure that failure of a hydraulic system will not result in loss of control.
- Multiple pressure sources: One or more engine-driven pumps plus one or more electric pumps for hydraulic pressure and the RAT as a backup source.
- Single hydraulic power supply failure does not influence flight control performance.
- Each engine drives dedicated pump(s), augmented by independently powered pumps (electric, pneumatic).
- No fluid transfer between systems to maintain integrity.
- System segregation.
- Route lines and locate components far apart to prevent single rotor or tire burst from affecting multiple systems.
- Hydraulic power cross supplies the surfaces on opposite sides.

- The left/right hydraulic power supply system should be connected when any one of the two fails.

Hydraulic actuator design philosophy is as follows:

- The actuator should be able to move the control surfaces with a following or opposing load while maintaining a rate of movement adequate for control purposes.
- The actuator should be able to hold the control surface at a required position with a load applied in either direction up to a defined maximum-load magnitude.
- The effect of the actuator frequency-response characteristics (gain and phase lag) on the low-frequency FCS loop stability margins should be minimized.
- Interaction with high-frequency (flexible aircraft) vibration modes should be minimized.
- Multiple actuators should be designed along the primary flight control axis.
- Multiple actuators on each surface should be powered by multiple hydraulic systems.
- Use a separately powered actuator or different types of actuators on one surface.
- Single actuator failures of primary control surfaces should not influence flight control performance.
- Split the surfaces of primary controls so as to increase the survivalability.
- Apply multiple control channels for critical surfaces.
- Each flight control needs multiple independent actuators or control surfaces.
- Left/right or upper/lower actuators should be reconfigured when any of them fails.

In the hydraulic system design, some principle failure modes and special failures should be considered. Principle failure modes include the following:

- Any single system or component failure (such as actuators, control spool housing, and valves)
- Any combination of failures (e.g., dual electrical or hydraulic system failures, or any single failure in combination with any probable hydraulic or electrical failure)
- Dormant failures of components or subsystems that only operate in emergencies
- Common mode failures/single failures that can affect multiple systems

Special failure cases consist of the following:

- One engine shuts down during takeoff—need to rapidly retract landing gear
- Engine rotor bursts—damage to and loss of multiple hydraulic systems
- Rejected takeoff—rapidly deploy thrust reversers, spoilers, and brakes

- All engines fail in flight—need to land safely without main hydraulic and electric power sources

In order to meet these requirements, it is necessary to design redundant hydraulic systems. Hydraulic system redundancy can be achieved by two primary means: multiple systems and multiple pressure sources within the same system.

The hydraulic power supply system and hydraulic actuation system should be verified by following a safety assessment tool before it is put into service.

- Failure modes, effects, and criticality analysis—computes failure rates and failure criticalities of individual components and systems by considering all failure modes
- Fault tree analysis—component failure rates and probabilities of various combinations of failure modes
- Markov analysis—computes the failure rates and criticality of various chains of events
- Common cause analysis—evaluates failures that can affect multiple components and systems

1.4.2 Requirement of Airworthiness [14,28,29]

According to the FAR Part 25—Airworthiness standards [14,30,3]: Transport Category Aircrafts, the aircraft may not have design features or details that experience has shown to be hazardous or unreliable. The requirements of Section 25.671 are presented in the following subsections.

1.4.2.1 Element Design [6]

Each element of the hydraulic system must be designed to

1. Withstand the proof pressure without permanent deformation that would prevent it from performing its intended functions and the ultimate pressure without rupture. The proof and ultimate pressures are defined in terms of the design operating pressure (DOP) as follows:

Element	Proof (×DOP)	Ultimate (×DOP)
Tube and fittings	1.5	3.0
High pressure (accumulator)	3.0	4.0
Low pressure (reservoirs)	1.5	3.0
Hoses	2.0	4.0
All other elements	1.5	2.0

2. Withstand the DOP in combination with limit structural loads that may be imposed without deformation that would prevent it from performing its intended function.

3. Withstand the DOP multiplied by a factor of 1.5 in combination with the ultimate structural load that can reasonably occur simultaneously without rupture.

4. Withstand the fatigue effects of all cyclic pressures, including transients, and associated externally induced loads, taking into account the consequences of element failure.

5. Perform as intended under all environmental conditions for which the aircraft is certified.

1.4.2.2 System Design [6]

Each hydraulic system must

1. Have means located at a flight crew station to indicate the appropriate system parameters, such as hydraulic system pressure, reservoir oil temperature, and oil level in the reservoir under normal conditions and malfunction.

2. Have means to ensure that system pressures, including transient pressures and pressures from fluid volumetric changes in elements that are likely to remain closed long enough for such changes to occur, are within the design capabilities of each element, such that they meet the requirements defined in Section 25.1435(a)(1)−(a)(5).

3. Have means to minimize the release of harmful or hazardous concentrations of hydraulic fluid or vapors into the crew and passenger compartments during flight.

4. Meet the applicable requirements of Sections 25.863, 25.1183, 25.1185, and 25.1189 if a flammable hydraulic fluid is used.

5. Be designed to use any suitable hydraulic fluid specified by the aircraft manufacturer, which must be identified by appropriate markings as required by Section 25.1541.

1.4.2.3 Tests [6]

Tests must be conducted on the hydraulic system(s) and/or subsystem(s) and elements. All internal and external influences must be taken into account to an extent necessary to evaluate their effects and to ensure reliable system and element functioning and integration. Failure or unacceptable deficiency of an element or system must be corrected and be sufficiently retested, where necessary.

1. The system(s), subsystem(s), or element(s) must be subjected to performance, fatigue, and endurance tests representative of aircraft ground and flight operations.

2. The complete system must be tested to determine proper functional performance and relation to the other systems, including simulation of relevant failure conditions, and to support or validate element design.

3. The complete hydraulic system(s) must be functionally tested on the aircraft in normal operation over the range of motion of all associated user systems. The test must be conducted at the system relief pressure or 1.25 times the DOP if a system pressure relief device is not part of the system design. Clearances between hydraulic system elements and other systems or structural elements must remain adequate, and there must be no detrimental effects.

1.4.3 Actuation System Performance Criteria [31]

Because the actuation system is used to drive the control surfaces, its performance should satisfy the specification of the actuation system. There are several types of requirements in actuation system design as presented in the following subsections.

1.4.3.1 Basic Data of an Actuation System

The basic data of an actuation system should be determined considering the aircraft design, aerodynamics, flying qualities, structure, flutter, and the following hydraulic and flight control parameters

- Specified oil supply pressure p_S and specified return oil pressure, p_R
- Maximum hinge moment of control surfaces, M_M
- Maximum angle of deflection of control surfaces, δ_{max}
- Maximum angular speed under zero load of control surfaces, $d\delta_{max}/dt$
- Hydraulic servo actuator-control surface transmission coefficient, K_δ
- Control surface around the shaft of the moment of inertia, J
- Control surface flutter suppression required by the hydraulic servo actuator-control surface natural frequency, ω_n
- The relative damping coefficient of the hydraulic servo actuator, $\zeta(\zeta = 0.03 - 0.08)$

1.4.3.2 The Static Requirements of the Hydraulic Servo Actuator [7]

The static performance requirements of the hydraulic servo actuator consist of stall load, maximum stroke, maximum rate capability, mechanical drive force, limit load of the hydraulic servo actuator, etc.

1.4.3.2.1 Stall Load

The stall load of a hydraulic actuator is the maximum force applied onto the main ram that can be supported by the hydraulic-supply pressure. The load can be in either the extend direction or retracted direction, and the criteria will apply equally for both. The stall load is a basic design parameter for the actuator and determines the required piston area for a given hydraulic-supply pressure available at the actuator. Stall load can be expressed as

$$P_{max} = C \cdot M_M \cdot K_\delta \tag{1.2}$$

where C is a constant and its value is $C = 1.10 \sim 1.15$. Theoretically, the hydraulic actuator provides the maximum power under maximum hinge moment when $C = 1.5$. However, it needs a maximum flow rate in this condition and would result in increase of the aircraft weight; therefore, it generally does not require that. Requirements on the actuator's load capability are usually defined as

- The minimum required output thrust (two systems operating with a defined pressure drop across each piston)
- The minimum single-system thrust (with a defined pressure drop across the pressurized piston, the other being bypassed)
- The maximum static-output thrust (two systems operating with a defined pressure drop across each piston).

These requirements are used to determine the size of the actuators within the set limitations. The first two will determine the minimum size (specifically piston area) to meet load and performance requirements, while the third requirement sets an upper limit on the size to prevent damage to the aircraft structure.

The magnitude of the design stall load is based on the maximum aerodynamic hinge moment predicted at any point in the flight envelope. In addition, it is required to ensure that there is sufficient excess capability in the actuator to provide driving forces under the most severe flight conditions. There are certain factors affecting the stall load, such as piston area, hydraulic-supply pressure, and the number of hydraulic systems operating. There are other factors influencing the stall load, including unequal piston areas on each side of the piston, force fighting in a tandem ram, and cross-piston leakage.

1.4.3.2.2 Maximum Stroke

The maximum stroke of the hydraulic servo valve can be calculated as

$$S_{max} = \frac{\delta_{max}}{K_\delta} + \Delta S \tag{1.3}$$

where ΔS is the adjustment range from 5 to 10 mm. Figure 1.31 shows the factors that should be considered when determining the maximum stroke of the hydraulic actuator.

1.4.3.2.3 Maximum Rate Capability [16]

Rate requirements are defined as a required rate, extending and retracting, for a given load and pressure differential across the piston. The required rates are

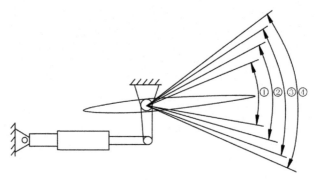

FIGURE 1.31   Factors related to the maximum stroke of a hydraulic actuator: ① control surface angle of deflection, ② adjustment range, ③ maximum mechanical stroke of the hydraulic actuator, and ④ minimum mechanical limit of the control surface.

usually defined at no-load conditions and approximately 60–70% of the stall load for two-system and single-system operation.

$$V_{\max} = \frac{d\delta_{\max}}{dt} \cdot \frac{1}{K_\delta} \tag{1.4}$$

The actuator supplier uses the rate requirements and the size of the actuator to determine the fluid flow rate needed and hence the necessary valve size.

The maximum rate at which the actuator main ram can be driven corresponds to the maximum opening of the valve ports. The maximum rate varies with the load and must be sufficient to move the aerodynamic control surface at a required speed. The hydraulic-supply pressure at the actuator must be ensured at the maximum rate capability and required performance. The factors affecting the maximum rate capability are steady load, supply and return pressures, the number of operating hydraulic systems, piston areas, main-valve port geometry, maximum main-valve displacement, cross-piston leakage, and valve-block pressure losses.

Figure 1.32 shows a typical plot of maximum rate capability versus external load up to the stall conditions for each direction of main-valve opening. The relationship between the maximum rate of a cross-piston leak and orifice diameter is also given in the plot. With a cross-piston leak, the main ram will sink against the demand direction when the actuator is applied at high steady load. This should be avoided in an actuation system design so as to ensure the valve port size and piston areas under all operating conditions.

1.4.3.2.4 Dead Zone [6]

The smallest amplitude of the step rate signal (or sinusoidal signal) that is required to make the actuator move from rest is defined as the dead zone. In general, the dead zone should be less than 0.1% of the specified input

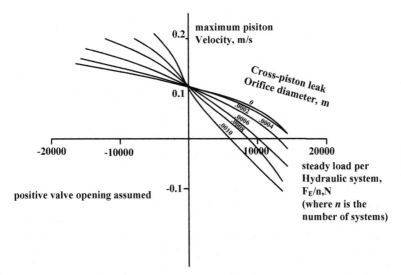

FIGURE 1.32   Typical actuator maximum rate versus external load.

command amplitude under zero load and rated pressure. The dead zone of the actuator is the result of the overlap of the orifice by the pilot valve spool.

1.4.3.2.5 Hysteresis

The hysteresis of the hydraulic actuator is defined as the maximum value of the output difference under the same input. Normally, the maximum hysteresis of 100% rated stroke amplitude is the 0.2% rated input command amplitude under zero load and an input command frequency of 0.004 Hz.

1.4.3.2.6 Force Coupling

The output difference among the redundant actuator output is required to be less than 10–20% of the rated force.

1.4.3.2.7 Mechanical Manipulating Force of the Hydraulic Actuator

According to the suppression pilot oscillation trend requirements and the driving rod of the authority for actuator motion response, the mechanical manipulating force of the hydraulic actuator is less than 10N.

1.4.3.2.8 Restricted Load of the Hydraulic Actuator

The restricted load is defined as 1.25 times the stall load of the hydraulic actuator. When the hydraulic actuator endures the limit load, the hydraulic actuator should not fail and should not undergo thus permanent deformation.

1.4.3.2.9 Maximum Load of Hydraulic Actuator

Maximum load is defined as 1.5 times the stall load of the hydraulic actuator. When the hydraulic actuator endures the maximum load, the hydraulic actuator could deform but should not loose bending stability, fail and result in external leakage.

1.4.3.3 The Dynamic Requirements of the Hydraulic Servo Actuator [7]

The dynamic requirements of the hydraulic actuator include the time response, frequency response, stability margin, and impedance characteristics.

1.4.3.3.1 Time Response

The requirements of actuator time response under a unit step are defined in the specification document as boundaries within which the time response must lie. Figure 1.33 shows the step response curve, in which four parameters should be considered:

- The maximum overshoot σ is the magnitude of the overshoot after the first crossing of the steady-state value (100%). This value is normally expressed as a percentage of the steady-state value of the controlled variable.
- The peak time t_p is the time required to reach the maximum overshoot.
- The settling time t_ε is the time for the controlled variable first to reach and thereafter remain within a prescribed percentage $\pm\varepsilon$ of the steady-state value. Common values of ε are 2%, 3%, or 5%.
- The rise time t_r is the time required to reach first the steady-state value (100%). It may also be defined as the time to reach the vicinity of the steady-state value, particularly for a response with no overshoot (e.g., the time between 10% and 90%). The 50% rise time $t_{r,50}$ is defined as the time to first reach 50% of the final value.

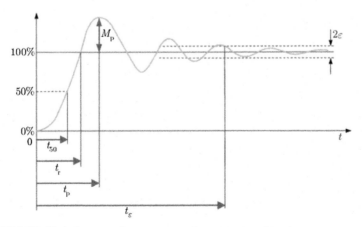

FIGURE 1.33 Typical actuator time response under step command input.

1.4.3.3.2 Frequency Response

Frequency response is the quantitative measure of the output spectrum of the actuator in response to a sine input. It is a measure of the magnitude and phase of the output as a function of frequency in comparison to the input.

The frequency response of a system can be measured by applying a test signal, for example:

- Applying an impulse to the system and measuring its response
- Sweeping a constant-amplitude pure tone through the bandwidth of interest and measuring the output level and phase shift relative to the input
- Applying a signal with a wide frequency spectrum and calculating the impulse response by deconvolution of the input signal and the output signal of the system.

The frequency response is characterized by the magnitude of the system's response, typically measured in decibels or as a decimal, and the phase, measured in radians or degrees, versus frequency in radians per second or Hertz. A Bode plot is a common way to describe the frequency response of a control system, Figure 1.34, in which a Bode magnitude plot expresses the magnitude of the frequency response, and a Bode phase plot expresses the phase lagshift. Both quantities are plotted against a horizontal axis given as the logarithm of frequency. The frequency bandwidth, time constant, and phase shift are indicators of dynamic performance of the actuator.

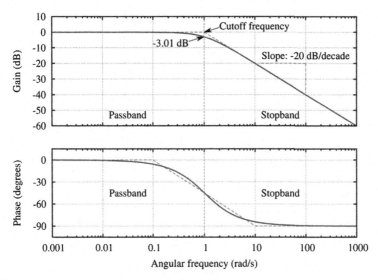

FIGURE 1.34 Typical Bode plot of a dynamic system.

Requirements of actuator frequency response are defined in the specification document as boundaries of the frequency response . The gain and phase-lag boundaries are applied to the response of the control-surface displacement to an input demand with representative inertia loading. Figure 1.35 shows an example of typical gain and phase boundaries applied to an FBW actuator. Bounds are placed on the maximum and minimum gain and on the maximum allowable phase lag, and a particular range of demand amplitudes is defined to encompass the linear range of operation of hydraulic actuator [7].

Frequency-response boundaries are defined to ensure that actuator effects on low-frequency modes are minimized with the gain of approximately 0 dB and the phase lag at a minimum while providing sufficient gain roll-off at high frequencies to reduce interaction with aircraft and control-surface structural

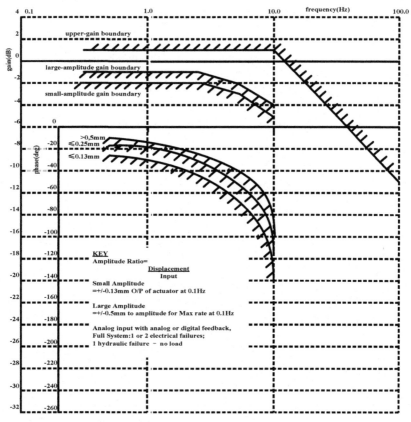

FIGURE 1.35   Typical frequency-response boundaries.

vibration modes. It should be noted that the gain and phase boundaries should be considered when control-surface structural modes are included.

The frequency response is a significant performance of the hydraulic actuator in which the bandwidth and phase lag are two main measures. The basic design goal of the actuator is to achieve the required performance for the specified range of frequencies and amplitudes. The upper-gain boundary is aimed primarily to avoid the presence of any excessive resonances. The lower gain boundaries apply for very small levels of input excitation with the amplitude changes. The factors affecting the low-amplitude response are nonlinear characteristics such as valve friction, valve laps, leakage, and hysteresis. The ability of the system to meet these specifications depends on the manufacturing accuracy of valve ports, ram friction, bearings, components, etc.

Phase boundaries express the maximum allowable phase lag at low or high amplitude. Satisfying the phase-lag criterion is important for the FCS, which represents the tracking accuracy of the actuator with respect to the command. No frequency response criteria are specified for very large amplitude inputs that will cause the valve to open to its full extent.

1.4.3.3.3 Dynamic Stiffness

Dynamic stiffness, or impedance, is the ability of the actuator to resist an external oscillatory load. Dynamic stiffness requirements are defined in the hydraulic actuator specification as boundaries within which the measured impedance must be located. Typical dynamic stiffness of a FBW aircraft actuator is shown in Figure 1.36 [7], where the constant force oscillatory

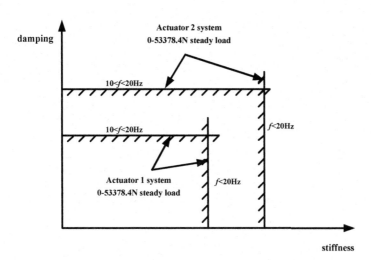

FIGURE 1.36 Typical impedance-response boundaries.

amplitudes are 4448.2 and 13,344.6N, the frequency range of the force of excitation is 10–50 Hz, and the fluid temperature is between 20 and 60 °C. The steady loads are

- Two systems: 0, 4448.2, 40,033.8, 53,378.4N
- One system: 0, 13,344.6, 26,689.2N

The dynamic stiffness of an actuator can be used to avoid control-surface flutter. Because the typical impedance is generally sufficient for the lower frequency range, the higher frequencies caused by flutter may be a critical issue for the hydraulic actuator. Normally, a hydraulic actuator can provide enough stiffness in conjunction with the output structure to the control-surface rotation mode so that the flutter-speed margins are met. There are ways to improve the dynamic stiffness of an actuator including

- Increase the area of the piston
- Set the leak hole between the cylinder cavities
- Add flutter damper
- Increase stiffness of the supporting structure

1.4.3.3.4 Failure Transients [7]

The requirements for failure transients are usually defined as boundaries on the ram-to-body displacement after the occurrence of the failures, such as electrical-lane failures, hard over failures, and hydraulic-supply failures. Figure 1.37 shows typical failure transient boundaries, in which class 1 failure boundaries apply to a first failure or a second failure if the first failed lane has been switched out and the class 2 boundaries apply to a first hydraulic failure and subsequent electrical failures. Failure transients are particularly affected by force fighting among the systems and main-valve pressure-gain characteristics.

1.4.4 Actuator Modeling [7]

During the dynamic design of an actuator, its mathematical model is established to determine its performance. An actuator mathematical model is used for simulation analysis to evaluate the response to step, ramp, sinusoidal, or other input demands. Transfer function analysis is performed to determine the gain and phase between input and output for a range of frequencies and amplitudes. To simplify the system model of the actuator, the first-order or second-order transfer functions are used to represent the hydraulic servo valve. For example, the flow rate of the valve is represented as a linear function of spool displacement using flow gain, as shown in Figure 1.38 [7]. There are two closed loops in this type of actuator. The inner loop measures the displacement of spool position and feeds it back to the servo amplifier to

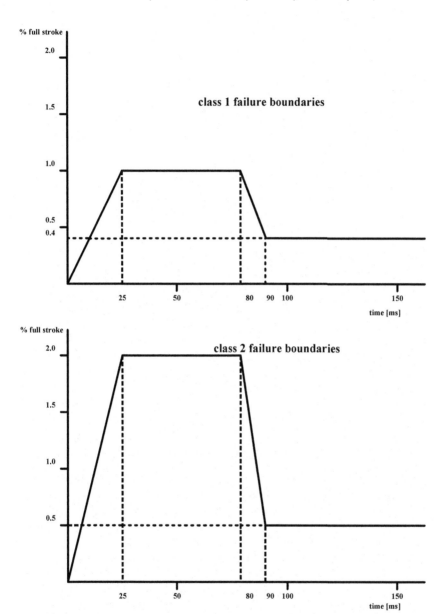

FIGURE 1.37 Typical failure transient boundaries [7].

achieve the high bandwidth. The outer loop obtains the cylinder output using LVDTs and provides feedback to the flight control computer to improve actuator performance. For analyzing the actuator performance, one needs to consider the effects of sampling time, computational delays, and

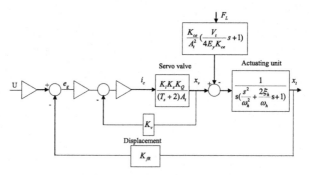

FIGURE 1.38 Block diagram of a typical actuation system, where K_i = gain of servo amplifier, K_s = gain of servo valve, K_Q = flow gain, T_s = time constant of servo valve, ω_h = inherent frequency of actuating unit, and ξ_h = hydraulic damping ratio.

nonlinearities, which influence the actuator stability, frequency response, and dither.

There are some nonlinear factors that need to be considered, such as nonlinear orifice flow and the fluid compressibility effects of the valve. The nonlinear model is used to assess pressure transients in stopping and starting an actuator, and the effects of saturation limits. Additional nonlinear factor includes the effects of friction or backlash of the valve.

1.4.5 Basic Parameters of Aircraft Hydraulic System

In hydraulic actuator design, some conventional parameters should be qualified in addition to the static and dynamic performance requirements.

1. Hydraulic-supply pressure: This will be determined by appropriate standards and the technology of the system. In today's commercial aircraft, 3000 psi (21 MPa) and 5000 psi (35 MPa) are widely used in hydraulic power supply systems. Typical commercial airline pressure is 3000 psi, while Airbus A380 has a 5000-psi hydraulic system.
2. Integrity: This requirement keeps aircraft flight safety under loss or degradation by adopting multiple independent sources of hydraulic power and the reversionary source, of power. Typical commercial aircraft adopts three or four independent hydraulic sources, whereas A380 utilizes a new dual-architecture system—two powered hydraulically and two powered electrically (2H+2E). Any one of these four systems can operate the aircraft, bringing the redundancy in flight control to a point unequaled on any other plane—commercial or military.
3. Flow rate: The rate of the demand, in angular or linear motion per second, or in liters per second, by to achieve the desired action.

4. Duty cycle: The ratio demand for energy compared with quiescent conditions. This will be high for continuously variable demands such as primary flight control actuation on an unstable aircraft, whereas it will be low for use as a source of energy for lowering and retraction of the undercarriage.
5. Emergency or reversionary use: Elements for the system that are intended to provide a source of power under emergency conditions for other power generation systems. An example of this is a hydraulic-powered electrical generator, which is used in reversionary devices to provide hydraulic-power in the event of main engine loss.
6. Heat load and dissipation: The amount of energy or heat load that the components of the system contribute to hydraulic fluid temperature.

Analysis of these aspects enables decisions to be made on the number and type of components required for the complete system. These components include the following:

- A source of energy—engine, auxiliary power unit, or RAT
- A reservoir
- A filter to maintain clean hydraulic fluid
- A multiple redundant distribution system—pipes, valves, shut-off cocks
- Pressure and temperature sensors
- A mechanism for hydraulic oil cooling
- A means of exercising demand—actuators, motors, pumps
- A means of storing energy, such as accumulator

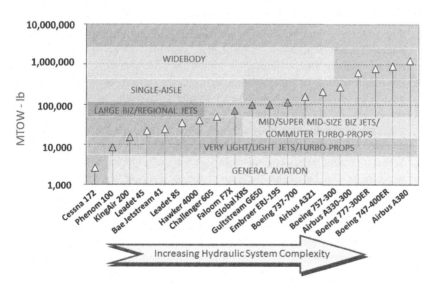

FIGURE 1.39   The development of an aircraft hydraulic system [10].

Figure 1.39 shows the relationship between size of the aircraft and complexity of the required hydraulic system of an aircraft hydraulic power supply system. A hydraulic system has many advantages, including the following:

- Effective and efficient method of power amplification
- Small control effort results in a large power output
- Precise control of load rate, position, and magnitude
- Infinitely variable rotary or linear motion control
- Adjustable limits/reversible direction/fast response
- Ability to simultaneously handle multiple loads
- Independently in parallel or sequenced in series
- Smooth, vibration-free power output
- Little impact from load variation
- Hydraulic fluid transmission medium
- Removes heat generated by internal losses
- Serves as a lubricant to increase component life

1.5 CONCLUSIONS

This chapter summarizes the development of hydraulic systems related to FCSs, and explains the interface between the FCS and the hydraulic system-actuation system. After introducing some typical actuation systems, the chapter provides design requirements of actuators and hydraulic systems. There are number of points that need to be considered in actuator design including

1. The stability of an FCS is the precondition of the aircraft, whereas the hydraulic servo actuator's operational stability is the most important requirement for FCS stability.
2. On the condition of meeting the performance requirements, the structure design of an actuator should be as simple as possible to ensure aircraft safety, reliability, and maintainability.
3. The pressure difference between both sides of the cylinder varies with the external load.
4. When hydraulic actuator-control surfaces are seriously unstable, the inner pressure of the hydraulic servo actuator of the actuator will exceed the hydraulic power supply pressure. When the external load of the actuator exceeds the stall load, the inner pressure is greater than the hydraulic power supply pressure; therefore, hydraulic system can adjust the flow characteristics of the actuator in the normal range.
5. The time constant is a very important parameter of the hydraulic actuator and is related to its tracking characteristics, stability, and impedance

characteristics. The time constant of a hydraulic actuator depends on the flow gain of the valve, feedback gain, and the piston area.

6. The FCS and aircraft structure parameters have a decisive influence on the actuator-control-surface characteristics. The greater the load inertia of the aircraft, the more unfavorable the aircraft stability.

7. The moment of inertia of the control surface around the shaft, the frequency of the actuator-control surface, the system damping coefficient behind the actuator, the maximum hinge moment of the control surface, and the maximum deflection of the control surface determine the stability boundary of the actuator-control surface.

8. The performance of the actuator directly influences the pilot-induced oscillation trend; therefore, the dynamic performance of the actuator is a very important issue in actuator design.

The mathematical model plays an important role in actuator design, especially in dynamic design. Linear and nonlinear factors should be considered in system design to meet the performance requirement.

REFERENCES

[1]   D. Briere, C. Favre, P. Traverse, Electrical Flight Controls, from Airbus A320/330/340 to Future Military Transport Aircraft: A Family of Fault-tolerant Systems, Aerospariale, CRC Press LLC, 2001. http://www.davi.ws/avionics/TheAvionicsHandbook_Cap_12.pdf.

[2]   J. Gao, Z. Jiao, P. Zhang, Aircraft Fly-by-Wire System and Active Control Technique, Beijing University of Aeronautics and Astronautics Press, 2010 (in Chinese).

[3]   DO-254: Design Assurance Guidance for Airborne Electronic Hardware.

[4]   MIL-F-9490D (USAF), Military Specification, Flight Control Systems−Design, Installation and Test of Piloted Aircraft, General Specification for, June 6, 1975.

[5]   T. Greetham, Evolution of Powered Aircraft Flight Controls, February 10, 2012.

[6]   I. Moir, A. Seabridge, Aircraft Systems: Mechanical, Electrical and Avionics Subsystems Integration, John Wiley & Sons, 2008.

[7]   R.W. Pratt, Flight Control Systems: Practical Issues in Design and Implementation, The Institution of Electrical Engineers & The American Institute of Aeronautics and Astronautics, 2000. What is a flight control system? wiseGEEK, http://www.wisegeek.com/what-is-a-flight-control-system.htm.

[8]   D. Crane, Dictionary of Aeronautical Terms, Aviation Supplies & Academics, 1997.

[9]   World Leader in Flight Control Systems and Critical Applications, Moog. http://www.moog.com/literature/Aircraft/Moog_AG_Aircraft_Capabilities_Brochure_Jun2012.pdf.

[10] P.A. Stricker, Aircraft Hydraulic System Design, Eaton Aerospace Hydraulic System Division Report, 2010, http://www.ieeems.org/Meetings/presentations/MS2-IEEE_Hyd_Systems_Presentation.ppt.

[11] http://aerospace.eaton.com/news.asp?articledate=06/01/03&NewsCommand=ViewMonth.

[12] http://www.tpub.com/content/aviation/14018/css/14018_178.htm.

[13] M. Jelali, A. Kroll, Hydraulic Servo System, Springer, 1982, ISBN 1852336927.

[14] T. Greetham, Evolution of Powered Aircraft Flight Control, MOOG report, 2012.

[15] D. Peter, Hydraulic Control Systems−Design and Analysis of Their Dynamics, Springer-Verlag Berlin Heidelberg, New York, 1981.

[16]   Y. Wang, Aicraft Flight Control Hydraulic Servo Actuator, Aeronautic Industry Press, Beijing, 2014.

[17]   J. Prokes, Hydraulic Mechanisms in Automation, Elsevier Scientific Publishing Company, Amsterdam, Oxford, New York, 1977.

[18]   N.D. Manring, Hydraulic Control Systems, John Wiley & Sons, 2005.

[19]   H.E. Merritt, Hydraulic Control Systems, John Wiley & Sons Inc., New York, London, Sydney, 1967.

[20]   R.B. Walters, G. Eng., F. I. Mech. E, Hydraulic and Electro-Hydraulic Control Systems, Elsevier Applied Science, London and New York, 1991.

[21]   C. Favre, Fly-by-wire for commercial aircraft—the Airbus experience, Int J Control (1993) special issue on "Aircraft Flight Control".

[22]   P. Traverse, D. Brière, J.J. Frayssignes, Architecture des commande de vol électriques Airbus, reconfiguration automatique et information equipage, INFAUTOM, 1994.

[23]   FAR/JAR 25: Airworthiness Standards: Transport Category Airplanes. Part 25 of "Code of Federal Regulations, Title 14, Aeronautics and Space," for the Federal Aviation Administration, and "Airworthiness, Joint Aviation Requirements—Large Aeroplane" for the Joint Aviation Authorities.

[24]   FAR Part 23: Airworthiness Standards for Normal, Utility, Acrobatic, and Commuter Category Aircrafts.

[25]   FAR Part 21: Certification Procedures for Products and Parts.

[26]   AC 25.1309-1A System Design and Analysis Advisory Circular, 1998.

[27]   ARP4761: Guidelines and Methods for Conducting the Safety Assessment Process on Civil Airborne Systems and Equipment.

[28]   ARP 4754: System Integration Requirements, Society of Automotive Engineers (SAE) and European Organization for Civil Aviation Electronics (EUROCAE), 1994.

[29]   ARP 4761: Guidelines and Tools for Conducting the Safety Assessment Process on Civil Airborne Systems and Equipment, Society of Automotive Engineers (SAE) and European Organization for Civil Aviation Electronics (EUROCAE), 1994.

[30]   AIR5005: Aerospace—Commercial Aircraft Hydraulic Systems.

[31]   DO178A, Software Considerations in Airborne Systems and Equipment Certification. Issue A. RTCA and European Organization for Civil Aviation Electronics (EUROCAE), 1985.

Chapter 2

Aircraft Hydraulic Systems

Chapter Outline

2.1 INTRODUCTION OF AIRCRAFT HYDRAULIC SYSTEMS

Hydraulic systems appeared in aircraft in the early 1930s to drive the retractable landing gear [1]. Aircraft hydraulic systems are currently used to transmit a large amount of power to move the flight control surfaces with little effort on the pilot's behalf. The system transmits and controls power from engine to flight control actuators and provides the force required to move large control surfaces under heavy aerodynamic loads. With the increasing power demand by the aircraft, the aircraft hydraulic system emerged as the solution for efficient transfer of the small low-energy movement by the pilot to the high-energy power output at aircraft surfaces. Using the hydraulic system, the pilot can move the control surfaces with increasing speeds and demands for maneuverability. The hydraulic system plays a very important role in modern aircraft, both military and civil, Figure 2.1 [2]. The hydraulic system today remains a most effective source of power for both primary and secondary flying controls and for landing gear and antiskid brake systems.

To ensure the safety of an aircraft, hydraulic system design requires that a single failure should not be allowed to place aircraft in hazard [3]. The high-level certification requirement of aviation regulations is that the aircraft should maintain control under all normal and anticipated fault conditions. Consequently, the aircraft adopts the multiple pumps, accumulators to store the hydraulic energy, and it deploys the monitoring devices identify and to isolate the faulty elements. A hydraulic system has many advantages as a power source for operating the units on aircraft, which are including [1,2]:

- Effective and efficient method of power amplification
  Small control effort results in a large power output

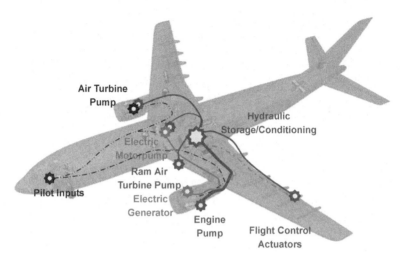

FIGURE 2.1   Aircraft hydraulic system.

- Precise control of load rate, position, and magnitude
  Infinitely variable rotary or linear-motion control
  Adjustable limits/reversible direction/fast response
- Ability to handle multiple loads simultaneously
  Independently in parallel or sequenced in series
- Smooth, vibration-free power output
  Little impact from load variation
- Hydraulic fluid transmission medium
  Removes heat generated by internal losses
  Serves as a lubricant to increase component life

Modern aircraft use fluids that meet the specifications of MIL-H-5606, MIL-H-83,282, and MIL-H-81,019 [1]. The general temperature of an aircraft hydraulic system varies from −65 to 295 °F. Conventional aircraft have adopted the hydraulic pressure of 3000 psi (21 MPa), whereas A380 has adopted a 5000-psi (35 MPa) hydraulic system. The hydraulic system applications within aircraft systems are shown in Figure 2.2.

Aircraft hydraulic control systems are used to control the position and/or speed of resisting aircraft loads, which require the power supply system and actuation system. An actuation system usually uses a hydraulic actuator (HA)—either a linear-motion hydraulic cylinder or a rotary-motion hydraulic motor. The actuator develops its force or torque by receiving liquid from a positive displacement pump at a relatively high pressure, usually from 3000 psi (21 MPa) to 5000 psi (35 MPa). A hydraulic servo-system is an arrangement of individual components, interconnected to provide a desired form of hydraulic transfer.

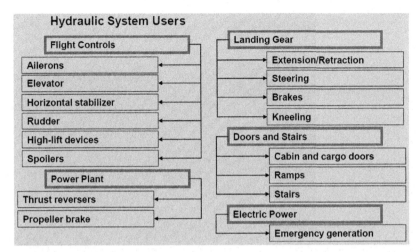

FIGURE 2.2 Aircraft hydraulic system users.

The disadvantages of hydraulic systems are as follows:

- They are heavy and require space (pump, pipes, etc.).
- The possibility of leakage, both internal and external, may cause the complete system to become inoperative.
- Contamination by foreign matter in the system can cause malfunction of the unit.

An aircraft hydraulic power supply system consists of an engine-driven pump (EDP), a power transfer unit (PTU), a hydraulic-driven generator, a hydraulic power package, a reservoir, an accumulator, a control valve, etc., as shown in Figure 2.3.

2.1.1 Basic Structure of a Hydraulic System

2.1.1.1 Single Hydraulic System

The basic structure of hydraulic systems is the single hydraulic system shown in Figure 2.4 [4], in which the main components consist of the pump, reservoir, accumulator, heat exchanger, filter, etc. The scope and size of a hydraulic system are determined by the requirement of the users of the system. These users need different power according to their application. In aircraft hydraulic system design, the pressure should be selected to move against the load according to the demand. The pressure of the hydraulic system varies from a couple 100 psi (or megapascals) in small aircraft to 5000 psi (35 MPa) in large aircraft.

Hydraulic fluids are used to transfer power, lubricate moving parts, self-off packing, and transfer heat. Pumps are used to pressurize the fluid, which

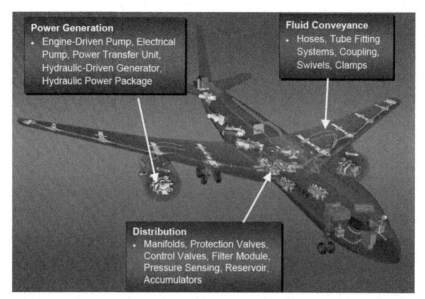

FIGURE 2.3   The components of an aircraft hydraulic power supply system.

moves through the hydraulic system and drives various elements, including actuators and motors. The hydraulic pumps can be classified according to the driving source: engine driven, electric motor driven, and ram air turbine (RAT) driven.

A hydraulic system consists of the following components [4]:

- Hydraulic power supply (pump, reservoir, relief valve, filter, cooler, etc.)
- Control elements (valves, sensor, etc.)
- Actuating elements (cylinder and/or motors)
- Other elements (pipelines, measuring devices, etc.)

The basic operation of a standard valve-controlled hydraulic system is briefly described below [4]:

- The pump converts the available (mechanical) power from the prime mover (electric or diesel motor) to hydraulic power at the actuator.
- Valves are used to control the direction of pump flow, the level of power produced, and the amount of fluid and pressure supplied to the actuator. A linear actuator (cylinder) or a rotary actuator (motor) converts the hydraulic power to usable mechanical power output at the required point.
- The hydraulic fluid provides direct transmission and control as well as lubrication of components, sealing in valves, and cooling of the system.

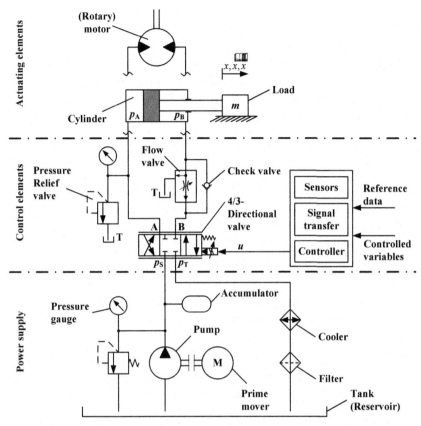

FIGURE 2.4   Basic structure of hydraulic systems.

- Connectors, which link the various system components, direct the power of the fluid under pressure and fluid-flow return to the tank (reservoir).
- Finally, hydraulic fluid storage and conditioning equipment ensure sufficient quality and quantity as well as cooling of the fluid.

The primary source of power in the aircraft is the engine, which also provides power to the hydraulic pump connected to the engine gearbox [1]. The pump causes a flow of hydraulic fluid at a certain pressure (ordinarily nominal 3000 psi (21 MPa)) through stainless steel pipes to various actuating devices. A reservoir ensures that sufficient fluid is available under all demand conditions. The shutoff valve (SOV) can shut down the fluid from the reservoir and pumps. The filter keeps the fluid clean. A nonreturn valve (NRV) ensures that the high-pressure oil flows only to the actuator. An accumulator provides the transient flow as demanded by the aircraft control system. Heat exchanger cools the return hydraulic fluid to a specified temperature. The servo valve (SV) is the control component of the actuator.

2.1.1.2 Dual-Redundant Hydraulic System

In hydraulic system design, the integrity should be considered to ensure the critical aircraft flight safety. In other words, the hydraulic system should provide the pressurized fluid under some loss or degradation. To meet the safety requirements, the hydraulic system adopts several independent sources of hydraulic power. The redundant hydraulic systems with multiple pumps and networks of pipes can ensure that single failures and leaks do not deplete the overall system of power. Figure 2.5 [2] shows the dual redundant hydraulic power supply system structure, in which P indicates pump and SV indicates the servo valve.

The degree of necessary redundancy is controlled by specifications and mandatory regulations issued by the national and international aircraft safety bodies. The requirements differ considerably between military and civil aircraft. Military aircraft usually adopt two independent hydraulic circuits whereas commercial aircraft invariably have three or more independent hydraulic circuits. In both types of aircraft, additional auxiliary power units and means of transferring power from one system to another are commonly implemented.

In Figure 2.5, the pump converts mechanical power into hydraulic energy (hydrostatic energy; i.e., flow and pressure), the filter filters the fluid contaminant, NRV indicates a nonreturn valve (viz. one-way valve), SOV indicates a shutoff valve, the accumulator can suddenly or intermittently release the stored pressure as per the requirement, the heat exchanger (also called the cooler) cools the hot fluid of return oil by using fuel heat sink cooling, the

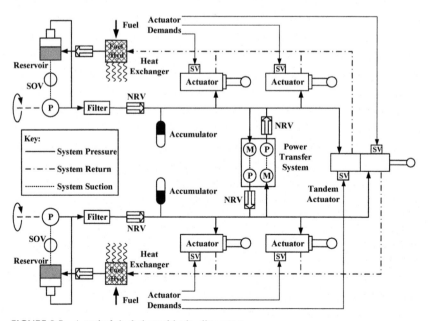

FIGURE 2.5   A typical dual-channel hydraulic system.

reservoir stores the fluid, the SV controls delivery of the fluid to hydraulic users, the actuator is the cylinder or motor that is responsible for moving the control surfaces, and the power transfer system provides an alternative source of power if one hydraulic power supply system fails.

In hydraulic system design, some design specifications should be complied to meet the system safety requirements, as follows [3]:

1. In case of failure, redundancy must be designed into the system. Because loss of hydraulics is a hazardous event, it is necessary to design a redundant hydraulic power supply system in aircraft. A manual control system requires less redundancy; fly-by-wire (FBW) requires more redundancy.
2. Safety assessment tools: The safety assessment process includes requirements generation and verification, which supports the aircraft development activities [5]. The process provides the methodology to evaluate the aircraft hydraulic system functions and determines the associated hazards of the system functions. The safety assessment tools include:
   a. Failure modes, effects, and criticality analysis (FMECA): Computes failure rates and failure criticalities of individual components and systems by considering all failure modes. FMECA also provides the severity of failure effects and finds out the weak link.
   b. Fault tree analysis: Identifies the failure events that could individually or collectively lead to the occurrence of the undesired top event.
   c. Markov analysis: Calculates the probability of the hydraulic system being in various states as a function of time.
   d. Common cause analysis: Identifies individual failure modes or external events that could lead to a catastrophic or hazardous failure condition.

   The detailed information will be shown in Chapter 3.

3. Principle failure modes considered:
   a. Single system or component failure
   b. Multiple system or component failures occurring simultaneously
   c. Dormant failures of components or subsystems that only operate in emergencies
   d. Common mode failures—Single failures that can affect multiple systems
4. Examples of failure cases to be considered:
   a. One engine shuts down during takeoff—need to rapidly retract landing gear
   b. Engine rotor bursts—damage to and loss of multiple hydraulic systems
   c. Aborted takeoff—rapidly deploy thrust reversers, spoilers, and brakes

All engines fail in flight—need to land safely without main hydraulic and electric power sources.

2.1.2 Hydraulic System of the Boeing Family [6,30,32]

Figure 2.6(b) shows the conventional hydraulic system structure in the Boeing family, in which three independent hydraulic systems are used: left, right, and center. Left and right hydraulic systems are independent with an EDP and an

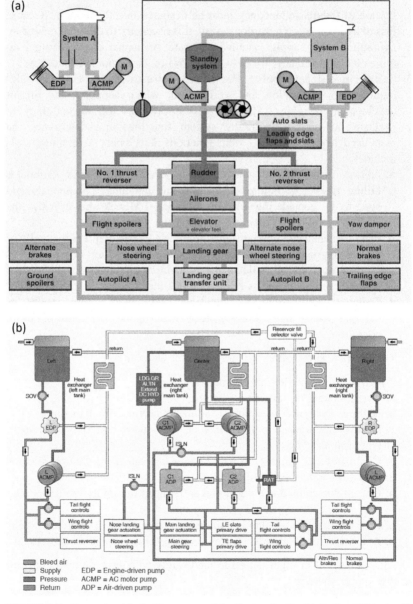

FIGURE 2.6   Conventional hydraulic system of the Boeing family. (a) Boeing 737 hydraulic system, (b) Boeing 777 hydraulic system.

alternating current motor pump (ACMP), which independently provide high-pressure fluid (3000 psi (21 MPa)) to the aircraft actuation system. If any of the two hydraulic systems were to fail, then the center hydraulic system is loaded to replace the faulty hydraulic system with two ACMPs. If an aircraft were to lose all of its hydraulic power, the RAT can ensure safe landing of the aircraft.

2.1.2.1 Boeing 737 Hydraulic System

For the Boeing 737, there are three independent hydraulic systems: System A, System B, and a standby system, Figure 2.6(a). The reservoirs are pressurized by bleed air through a pressurization module. The standby reservoir is connected to the System B reservoir for pressurization and servicing. System A and B hydraulic systems have an EDP and an ACMP. The EDP of System A is installed on the No. 1 engine and the one of System B is installed on the No. 2 engine. The ACMPs are controlled by a switch on the flight deck. The hydraulic fluid lubricates and cools the pumps and returns to the reservoir through a heat exchanger. The heat exchanger for System A is installed in the main fuel tank No. 1 and the heat exchanger for System B is installed in the main fuel tank No. 2. On the ground, the electric motor-driven pump operates to provide the power. Pressure switches are located in the EDP and ACMP output lines and send a signal to illuminate the LOW PRESSURE light if EDP pressure is low. The related system pressure transmitter sends the combined pressure of EDP and ACMP to the hydraulic system pressure indicator.

The PTU is used to supply the additional volume of hydraulic fluid needed to operate the slats and leading edge flaps at the normal rate when the EDP of System B fails. The PTU unit consists of a hydraulic motor and hydraulic pump that are connected through a shaft. PTU uses System A pressure to drive a hydraulic motor, and the hydraulic motor connects the hydraulic pump that can draw fluid from the System B reservoir. The PTU can only transfer power and cannot transfer fluid. The PTU operates automatically when the EDP of System B pressure drops below limits or the flaps are less than 15° but not up.

The standby hydraulic system is provided as a backup if System A and/or B pressure is lost. The standby system can be activated manually or automatically and uses a single electric ACMP to power thrust reversers, the rudder, leading edge flaps and slats (extend only), and the standby yaw damper.

When an aircraft has lost all hydraulic power, the RAT can ensure the safe landing of the aircraft.

2.1.2.2 Boeing 777 Hydraulic System

Let us take Boeing 777, Figure 2.6(b), as an example to demonstrate the operational principle of a hydraulic system. The aircraft is equipped with three independent hydraulic systems: left, center, and right systems. The three systems provide hydraulic fluid at the rated pressure of 3000 psi (21 MPa) to operate the flight controls, flap systems, actuators, landing gear, and brakes. Left and right hydraulic systems are identical and deliver the primary hydraulic power by two EDPs and are supplemented by two on-demand ACMPs. The left and right hydraulic systems each have a demand pump that is driven by an electric motor. The demand pumps provide supplementary hydraulic power for periods of high-system demand. The demand pumps also provide a backup hydraulic power source for EDPs. The pumps are controlled by demand L and R pump selectors. In the AUTO position, the left and right demand pumps operate for takeoff, landing, and when system or EDP pressure is low. In the ON position, the demand pump continuously runs. The center hydraulic system provides hydraulic fluid by two ACMPs and is supplemented by two on-demand air turbine-driven pumps (ADPs). The C1 and C2 pump switches control pump operation. The center hydraulic system powers flight controls, leading edge slats, trailing edge flaps, landing gear actuation, engine thrust reversers, etc. In case of emergency, the RAT is automatically deployed and drives a variable-displacement inline pump to generate the hydraulic power for landing safely when an aircraft loses all hydraulic systems. The RAT pump provides fluid to the center hydraulic system.

On the ground, a single ground power source including the auxiliary power unit (APU) and C1 is used whereas the C2 pump is not run. The pump will not be load shed if one engine generator is operating or the external power and APU generator are operating. In flight, the C2 pump may be load shed by the electrical load management system when all other electric pumps are running, there is a single source of electrical power, or the generator capability is exceeded. The pump will start automatically when the conditions that shed the pump no longer exist. The function of the ADP in the center hydraulic system is similar to the left or right demand pump. The pumps are controlled by demand C1 and C2 pump selectors. In the AUTO position, the demand pump operates when system or EDP pressure is low or system logic anticipates a large demand. In the ON position, the demand pump continuously runs. Selecting both demand pumps ON results in the operation of only pump C1. C1 and C2 cannot operate simultaneously when ON is selected.

The RAT pump automatically provides fluid to the center hydraulic system when both engines have failed and the center system pressure is low, both AC transfer buses are unpowered, or all three hydraulic systems' pressure are low. The RAT can be deployed manually by pushing the RAM AIR TURBINE switch. The hot battery or APU battery bus must be powered. Once deployed, the RAT cannot be stowed in flight.

2.1.3 Hydraulic System of the Airbus Family

The hydraulic system of the Airbus family integrates two processes: the initial stage adopts multiple hydraulic systems similar to the Boeing family, whereas

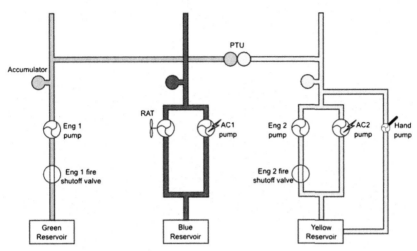

FIGURE 2.7   A320 family hydraulic system.

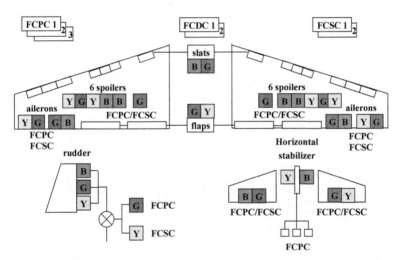

FIGURE 2.8   The layout of the actuation system related to the hydraulic system of A340.

the latter stage utilizes the dissimilar redundant electrohydraulic power supply systems such as Airbus 380 (two hydraulic power supply systems and two electrical power supply systems (2H+2E)). The hydraulic systems in A320, A340, and A380 aircraft are shown in Figure 2.7−2.9 [1].

2.1.3.1 A320 Aircraft Hydraulic System [1,7]

There are three continuously operating hydraulic systems in A320: green, yellow, and blue, as shown in Figure 2.7. Three hydraulic systems are

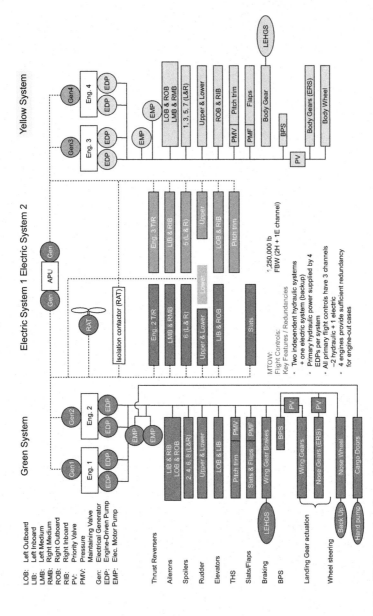

FIGURE 2.9   Flight control surfaces of A380 [1].

segregated. Each hydraulic system has its own pressurized reservoir to prevent cavitation. Hydraulic fluid cannot be transferred from one system to another. The common failures of reservoir monitoring are low fluid level, low hydraulic pressure, and overheating. The green and yellow systems are pressurized by an EDP, which can deliver hydraulic fluid at 37 gallons per minute (gpm) or 140 L/min to the actuation system. The yellow hydraulic system may also be pressurized by the ACMP, which can be powered by either AC2 or external power when the engines are stopped and can deliver hydraulic fluid at 6.1 gpm or 23 L/min fluid for ground service operation. It operates automatically to partially pressurize the yellow system when an increase in flow rate is required during flight (e.g., the cargo doors move). The yellow hydraulic system also has a hand pump for crew members to pressurize the cargo door operation when no electrical power is available on the ground. The blue system is pressurized by an ACMP (AC1 in Figure 2.7), which operates whenever AC power is available and can deliver 6.1-gpm or 24-L/min flow. In an emergency situation, the blue system may also be pressurized by the RAT, which can deploy automatically if both AC BUS 1 and AC BUS 2 are lost and can provide 20.6 gpm or 78 L/min at 2175 psi (15.225 MPa).

The EDPs and the electric pumps deliver hydraulic fluid at 3000 psi (21 MPa) hydraulic pressure, whereas the RAT generates 2175 psi (15.225 MPa) pressure. The main failures of the pump monitoring system are low output pressure and overheating. The main failure of the electric pump monitoring system is overheating. A bidirectional PTU enables the yellow system to pressurize the green system when the pressure difference between the green system and the yellow system exceeds the set value and vice versa. It automatically activates when the differential pressure between the green and the yellow system is greater than 500 psi (3.5 MPa). On the ground, an ACMP is used to pressurize the yellow hydraulic system, whereas the PTU allows for pressurization of the green hydraulic system. The RAT extends automatically in the flight when both engines and APU fail. In the event of an engine fire, a SOV in the suction line between the EDP and the reservoir will be closed. An accumulator is provided for each system to help maintain constant pressure and enough flow during transient demands. Each system has a priority valve to cut off heavy users (flaps, slats, gear, emergency generator) if the system pressure gets too low to operate the flight controls.

All of the actuation systems described above are conventional HAs, which are connected to the central hydraulic power supply system of the aircraft through pipes. The HA consists of the SV, cylinder, and displacement sensor. The SV can direct the hydraulic pressure to move the actuator shaft, which is connected to the control surface (aileron, elevator, rudder, spoiler, etc.).

2.1.3.2 A340 Aircraft Hydraulic System [8]

The hydraulic system design philosophy of A340 is similar to that of A320, which has three fully independent hydraulic systems (green, blue, and yellow),

Figure 2.8. There are four engines in the A340. Each engine has one engine-driven hydraulic pump. Engines 1 and 4 have a pump that pressurize the green system, Engine 2 drives the pump for the blue system, and Engine 3 drives the pump pressurizing the yellow system. In addition, there are three ACMPs in A340: two pumps are for the blue system and one pump is the backup for the yellow system. The pressure of the hydraulic system is 3000 psi (21 MPa). The green and yellow systems are each powered by an EDP, and the blue system is powered by an ACMP. When the engine operates, three hydraulic systems automatically provide high pressurized fluid. Two EDPs connect their corresponding engines by means of an additional gearbox. When the primary hydraulic systems cannot provide the power to the users, the auxiliary hydraulic system of the aircraft supplies the power. The auxiliary hydraulic systems include the PTU, an ACMP for the yellow system, and the RAT. The PTU can cross-pressurize between yellow and green systems without transfer of fluid when the pressure difference between the two systems is greater than 500 psi (3.5 MPa). PTU operates automatically and transfers the pressure from the high-pressure system to the low-pressure system. The RAT of the yellow system can provide the power to the actuation system when two engines fail under emergency conditions. Even if the aircraft were to lose the entire AC bus, RAT can automatically extend operation. The ACMP of the yellow system can provide hydraulic power to the yellow system when its engine or the EDP fails. On the ground, the ACMP of the yellow system can be used to provide the power.

The layout of the actuation system related to the hydraulic system of A340 is shown in Figure 2.8.

In Figure 2.8, the green hydraulic system provides the high-pressure fluid to the elevator, aileron, rudder, slat, flap, and spoiler. The yellow hydraulic system pressurizes the right elevator, aileron, rudder, flap, and spoiler. The blue hydraulic system provides the fluid to drive the left elevator, aileron, slat, and spoiler. All of the actuators used in the Airbus 340 are HAs. Only the spoiler usage on A340 differs from that on A320; namely, there are six pairs of spoilers in A340, whereas there are five in A320. The hydraulic system in A340 monitors its system pressure, pump pressure, reservoir level, and reservoir temperature in real time. In the case of low pressure, a priority and pressure-maintaining valve preserves pressure for the brakes, flight controls, and thrust reversers whereas other controls will be lost.

In the early 1990s, Airbus designed a new type of actuator, which decentralizes the hydraulic system architecture and uses the aircraft electrical network [11]. The new design is an electrohydrostatic actuator (EHA) with a self-contained electrical hydraulic pump, reservoir, and accumulator. The EHA can generate hydraulic power and move the control surfaces without the pipe network and the central hydraulic power supply system. The EHA was initially used in the aileron of A320 MSN 1, and it operated for more than 500 h on the A320 Iron Bird until the late 1990s. Finally, the EHA was developed and evaluated in the A330/A340 inner aileron from 1998 to 1999.

Therefore, A320 and A340 contributed to the widespread use of the EHA in the Airbus family.

2.1.3.3 A380 Hydraulic System [1]

The hydraulic system design in A380 aircraft adopts a totally different philosophy. The aircraft implemented two hydraulic power supply systems (2H) and two electrical power supply systems (2E). This philosophy has been widely used by Airbus over the past 20 years. Figure 2.9 shows the A380 flight control actuation configuration and Figure 2.10 shows the 2H+2E architecture. There are two independent hydraulic systems (green and yellow system) and one electric system backup in A380, where E1 and E2 replace the third hydraulic system. A380 takes the lead in the use of the high-pressure (5000 psi (35 MPa)) hydraulic system. Normally, two hydraulic power supply systems provide the high-pressure fluid to the primary flight control system, landing gear, nose wheel steering system, and other hydraulic users. In addition, there are four electrical motor pumps (EMPs) and the related electric control and protection systems.

On A380 aircraft, many of the actuators are powered by the green, blue, and yellow hydraulic systems. Others are powered electrical signal, that is EHAs, as backup. The two outboard aileron surfaces and six spoiler surfaces of each wing are driven by HAs—yellow or green system. The mid- and inboard aileron surfaces and the outboard elevator surfaces are powered by HAs and EHAs. The primary actuation system is the HA; the EHA will replace the HA in the case of HA failure. The two spoiler surfaces and rudder are powered by the electrical backup hydraulic actuators, which combine the features of HAs and EHAs. The trimmable horizontal stabilizer actuator is powered independently from the green and yellow hydraulic systems and from the electrical power supply.

In A380 aircraft, the HAs are normally active whereas the electrically powered actuators are normally standby. The EHA becomes operative in the event of a failure of the normal, hydraulically supplied and control lane.

In A380 aircraft, the actuation system receives the hydraulic input from the centralized hydraulic power supply system (green or yellow) and the electrical channels (E1 or E2 or exceptionally E3 AC essential (RAT)). Their drive mode is divided into two kinds: the primary drive depends on the hydraulic supply system (i.e., HA mode); while the backup drive relies on the electrical power supply (i.e., EHA mode).

- Normal mode—HA mode: In the normal condition, the HA receives the demand from the FBW computer, and the SV moderates the hydraulic supply from the green or yellow hydraulic system to drive the actuator movement.
- Backup mode—EHA mode: In the backup mode, the EHA receives the demand from the FBW computer, and the electrical motor operates with the electrical power from the aircraft AC electrical system. The electrical motor drives the fixed displacement hydraulic pump to move the cylinder.

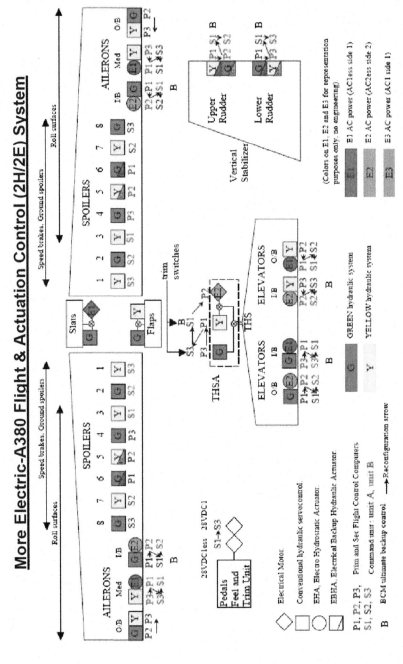

FIGURE 2.10 "2H/2E" structure of A380 [1].

The A380 philosophy is more toward the electric flight control actuation concept by eliminating one hydraulic system and replacing it with a set of electrical power supplies. This approach does not have a does it have positive effect? on the probability of losing the flight control actuation system. The selected power source distribution, identified as "2H/2E," features two hydraulic systems called green and yellow and two electric systems called E1 and E2. With this kind of dissimilar power supply design, A380 aircraft is more reliable and safer than other types of aircraft designs.

Most aircraft in current service have hydraulically powered actuators. The high safety requirement of the critical system demands that the hydraulic system has multiple pumps, accumulators to store energy, and methods for isolating leaks. The hydraulic system has several special features, including high-power density, flexible installation, oil lubrication, and robustness to overload, which make them very attractive for application in aircraft flight controls. The following sections provide an overview of the main components in aircraft hydraulic systems.

2.2 BASIC PARAMETERS OF AN AIRCRAFT HYDRAULIC SYSTEM

According to the previous analysis, most aircraft used today need hydraulic power for many tasks. Therefore, the most important thing in hydraulic system design is to meet the flight safety requirements and to accomplish the required functions. The functions of the hydraulic system include primary flight controls (elevators, rudders, ailerons), secondary flight controls (flaps, slats, spoilers), and utility systems (landing gear, antiskid brake, cargo doors, etc.). Because the primary flight controls are critical to flight safety, not a single failure is allowed to even momentarily interrupt their operation. The principle of the aircraft hydraulic system is to provide the high-pressure fluid to the actuation system to drive the control surfaces and other users. There are several critical parameters that need to be considered when designing aircraft hydraulic systems. Those are discussed in the following sections.

2.2.1 Hydraulic Fluid

The hydraulic fluid is a physical medium that can transmit power in a certain temperature range and at a maximum flow rate. There are currently differences in hydraulic fluid specifications between military aircraft and commercial aircraft. The hydraulic fluid requirements for military aircraft are covered under DTD 585 and MIL-H-5606. Commercial aircraft use the fire-resistant phosphate ester fluids under Solutia Skydrol LD-4, Skydrol 500B-4, or Skydrol 5. These fluids are not fireproof; because of the combination of fluid spray and hot surfaces they ignite and burn.

2.2.2 Hydraulic Pressure

When hydraulics were introduced into aircraft in the 1930s and 1940s, they operated at 1500 psi (10.5 MPa). In the 1950s, 3000 psi (21 MPa) became the standard. Over the past several decades, aircraft manufacturers chose 3000 psi (21 MPa) or 4000 psi (28 MPa) as the operating fluid pressure that can keep the minimum weight under different working profiles. Many studies have been undertaken to increase the standard operational pressure so as to further reduce the system component mass and volume. As a result, the pressure targets have varied from 5000 psi (35 MPa) to 8000 psi (56 MPa) in recent years. In practical application, different operational profiles have different fluid pressure requirements; therefore, it is necessary to adopt the variable hydraulic pressure system for saving energy (discussed in Chapter 4). The conventional values of aircraft hydraulic pressure are shown in Table 2.1.

There are three kinds of pressure in a hydraulic system: rated pressure, overpressure, and maximum full-flow pressure. The rated pressure is the pressure against which the pump is required to operate continuously at rated temperature and rated speed. The overpressure is the maximum transient pressure—approximately 125% of the rated pressure. The conventional overpressure values are listed below:

- 1500-psi system: the maximum pressure is 1875 psi
- 3000-psi system: the maximum pressure is 3750 psi
- 4000-psi system: the maximum pressure is 4850 psi
- 5000-psi system: the maximum pressure is 5850 psi
- 8000-psi system: the maximum pressure is 8850 psi

The maximum full-flow pressure is the maximum discharge pressure at which the rated flow can be developed at rated temperature, rated speed, and rated inlet pressure. Its value shall be no less than 95% of the rated discharged pressure, unless indicated differently in the detailed specification.

TABLE 2.1 Conventional Hydraulic Pressures Used in Aircraft

psi	MPa
1500	10.5
3000	21
4000	28
5000	35
8000	56

TABLE 2.2 Conventional Hydraulic Fluid Temperature

Hydraulic System	Maximum System Temperature °F (°C)	Rated Temperature at Pump Inlet				
		psi 1500 MPa 10.5 °F (°C)	3000 21 °F (°C)	4000 28 °F (°C)	5000 35 °F (°C)	8000 56 °F (°C)
Type I	160(70)	110(45)	110(45)	105(40)	95(35)	75(25)
Type II	275(135)	225(110)	225(110)	220(105)	210(100)	190(90)
Type III	450(232)	390(200)	390(200)	385(195)	375(190)	355(180)

2.2.3 Fluid Temperature [1]

Aircraft fluid temperature is limited by the fluid used. For many years, the maximum temperature of hydraulic fluid (DTD 585) was limited to approximately 130 °C, and the hydraulic components and seals were qualified accordingly. Recently, some commercial aircraft increased this limit to 200 °C, and other suitable hydraulic fluids have been used. The phosphate ester-based fluid used in the current commercial aircraft has good fireproofing performance, but it easily hydrolyzes and oxidizes. Another issue is that the fluid viscosity will decrease as the temperature increases. This will cause a reduction in lubrication and potential damage to the connected actuator and motor. The conventional values of rated fluid temperatures are listed in Table 2.2.

2.2.4 Fluid Flow Rate [1]

The flow rate is determined by the aerodynamic load and load movement velocity. The flow rate needs to meet the requirements of the hydraulic power supply system, SV, and actuator. Once the hydraulic pressure is chosen, the designer needs to consider the pressure drop from pump to reservoir under the full-flow condition, which is usually approximately 20–25% of the rated pressure. When designing an HA under static loading conditions, the piston area and distance of travel can be determined according to the aerodynamic loads. When designing an HA for dynamic loading conditions, the flow rate of the SV can be determined under the required maximum movement velocity. The flow rate should provide certain allowance for normal leakage. The sum of these will determine the maximum flow rate demanded of the system.

Normally, different flight profiles have different flow rate requirements (e.g., takeoff, cruise flight, etc.). The maximum flow rate is commonly needed only for a very short duration. Hence, it is not necessary to design a hydraulic system with this value of fluid flow. The common approach to dealing with a temporary increase in flow demand is to include an accumulator to augment the available flow.

2.2.5 Hydraulic Pipes [1]

Hydraulic pipes are widely used to connect the hydraulic power supply to the actuation system. When the system architecture is defined for all aircraft systems using hydraulic power, the next step is to design the pipe layout in the aircraft. This layout should consider the need to separate pipes to avoid the common mode failures as a result of accidental damage. When the hydraulic pipe network and layout are designed, the lengths and the weights of the pipes can be determined. In addition, the designer should consider the pressure drop of pressure pipes, return pipes, and components. Once the pipe lengths, flow rates, and permissible pressure are known, pipe diameters can be calculated.

2.2.6 Pressure Pulsation [1]

Pressure pulsations are the oscillations of the discharge pressure, occurring nominally during steady operating conditions, at a frequency equal to or higher than the pump drive shaft speed. In general, the amplitude of pressure pulsations shall not exceed 5% of the rated pressure under any condition or a pressure band specified by the detailed pump specification. A380 aircraft requires that the pressure pulsation of a hydraulic pump be less than or equal to $\pm 1\%$. Eaton Corporation designed the built-in attenuator and hydraulic pump with 11 pistons to solve the above difficulties. The hydraulic pump should be tested in the circuit that simulates the actual system in which the pump is to be installed, as defined in the detailed pump specification. The system volume shall be simulated using tubing of the discharge line diameter. A tubing line length for which the natural frequency is resonant with the pulsation frequency must be avoided.

2.3 MAIN COMPONENTS OF THE AIRCRAFT HYDRAULIC SYSTEM

2.3.1 Aircraft Hydraulic Pump

2.3.1.1 Engine-Driven Pump [9,10]

The hydraulic system uses an EDP to supply the pressurized hydraulic fluid for the flight control system and utility systems; therefore, EDP is the most important component in the hydraulic system. The typical EDP is a pressure-compensated, variable displacement axial piston pump capable of delivering a variable volume of fluid to maintain pressure in a hydraulic system. The EDP

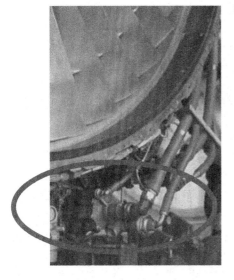

FIGURE 2.11   EDP installation position.

is normally mounted to the engine gearbox and driven by a splined input shaft, Figure 2.11.

Figure 2.12 shows the outside view and cross-sectional view of the EDP, in which nine pistons are nested in a circular array within the cylinder barrel. The external interface of the EDP includes the drive shaft, mounting flange, inlet, discharge, and case drain outlet. The inside of the EDP includes the cylinder barrel, piston, port plate, hanger, transmission shaft, and swash plate. The cylinder barrel is held tightly against the port plate using the compressed force of the spring. A thin film of oil separates the port plate from the cylinder barrel, which forms a hydrodynamic bearing between the cylinder barrel and port plate under normal conditions. The cylinder barrel is connected to the shaft through a set of splines that run parallel to the shaft. The port plate has two kidney-shaped ports that connect the inlet port and drain-out port, respectively. Fluid enters and leaves the barrel through the kidney-shaped slots in the port cap. A ball-and-socket joint connects the base of each piston to a slipper. The slippers are kept in reasonable contact with the swash plate by a retainer, where a hydrostatic and hydrodynamic bearing surface separates the slippers from the swash plate. The swash plate angle varies with time to create variable displacement of the pump.

The operational principle of the hydraulic pump is shown in Figure 2.13 [1].

As the EDP rotates, the pistons reciprocate within their bores, and in the process, they receive the low-pressure fluid from the "inlet port" and discharge it at high pressure through the "outlet port." By changing the swash plate angle, the piston stroke can vary and the flow rate of the pump will change. The pressure of the EDP is dependent on the external load. Axial forces on the

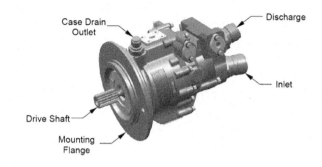

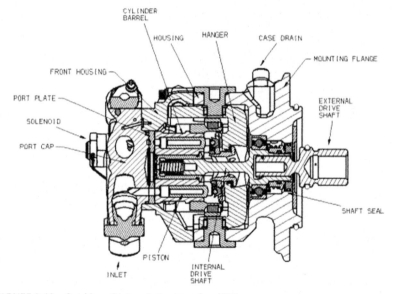

FIGURE 2.12   Outside and internal structure of an EDP.

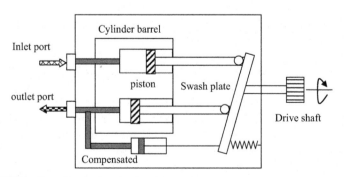

FIGURE 2.13   Operational principle of a piston pump.

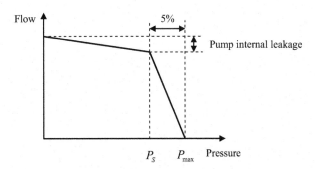

FIGURE 2.14   Flow rate versus pressure of an EDP.

piston/shoe subassembly are balanced by porting pressurized fluid through the piston neck to a hydrostatic balance area under the piston shoe. The barrel slots are designed with a hydrostatic balance. The cylinder barrel is allowed to align itself with the port cap through the arrangement of the drive shaft and cylinder barrel bearings. Figure 2.14 [1], shows the pump's characteristic curve, pressure versus flow rate, where P_S is the pump's rated pressure and P_{max} is the pump's maximum pressure. Under normal conditions, EDP can maintain the rated pressure within 5% of the nominal value by rapidly adjusting the swash plate from low flow to high flow. EDPs are designed to be sensitive to the outlet pressure and to be able to feed back this signal to the plate carrying the reciprocating piston until the swash plate returns to the preset angle.

The EDP discharge pressure is controlled by the compensator. The pressure compensator maintains the delivery pressure by adjusting the swash plate angle and resulting discharge flow in response to changes in the system pressure. The compensator valve performs integration by metering the volume of fluid to the stroking piston that is proportional to the change in pump output. In turn, the stroking piston controls the swash plate to determine the amount of fluid expelled, thus providing the variable delivery required.

Table 2.3 shows different types of EDPs manufactured by Parker Corporation.

Recent design of EDPs for commercial aircraft has several improved features, such as compactness, lightweight, increased durability, and high efficiency. The pump timing and displacement controls provide for low-pressure ripple and smooth response to rapid changes in flow demand.

Different EDPs have different characteristics. The basic parameters of EDPs include

- Rated operation pressure: 3000 psi (21 MPa)
- Proof pressure: 4500 psi (31.5 MPa)
- Burst pressure: 7500 psi (52.5 MPa)
- Displacement: for example, 0.96 cipr (cubic inches per revolution)

TABLE 2.3 EDP Manufactured by Parker [9]

Model Number	Maximum Displacement (in³/rev)	Normal Operating Pressure (psi)	AS595 Maximum Recommended Speed (rpm)	Maximum Output Flow (gpm)	Approximate Dry Weight (lbs)	Approximate Envelope Length (in)	Height (in)	Width (in)
AP05VC	0.09	3000	13,060	4.8	2.4	3.6	3.3	3
AP05V	0.15	3000	11,034	6.8	2.6	3.9	4.3	3.8
AP1V	0.31	3000	8684	11.1	7.0	6.5	5	4.3
AP2V	0.42	3000	7856	13.6	8.0	6.5	5	5
AP3V	0.52	3000	7321	15.7	9.0	7	5.7	4.6
AP4V	0.65	3000	6801	18.2	10.5	7	6	6
AP5V	0.82	3000	6299	21.2	11.5	7	6	6
AP6VSC	0.97	3000	5960	23.8	12.0	7.2	6	6
AP8V	1.35	3000	5344	29.7	16.0	7.3	6	6
AP9VM	1.20	5000	5085	25.1	21.5	9.5	6.7	6.7
AP10VC	1.60	3000	5052	33.2	15.3	9.4	6	6
AP12V	2.02	3000	4677	38.9	18.5	7.2	7.5	6
AP15V	2.40	3000	4420	43.6	25.0	10.4	7.5	7
AP19V	3.00	3000	4106	50.7	27.5	12.8	9	9
AP27V	4.30	3000	3646	64.5	29.8	10.3	8.6	6.8
AP36V	5.50	3000	3361	76.0	47.0	9.8	8.5	8.5
AP20VM	3.05	5000	3737	46.9	41.0	13	9.5	8.5

- Full-flow pressure: 2850 psi (20 MPa)
- Rated output flow: 17.0 psi
- Rated inlet pressure: 39 psi
- Rated case drain pressure: 80 psi
- Case drain flow: 0.3 gpm at 2913–4310 rpm
- Rated continuous inlet fluid: 20 °F (6.67 °C) to 225 °F (107.2 °C)
- Dynamic shaft seal leakage: 1 drop/2 min maximum
- Static shaft seal leakage: 1 drop/10 min maximum

The theoretical displacement of the pump is the amount of change of the working volume of all plunger cavities when the cylinder barrel rotates a full circle; that is,

$$Q_l = \frac{\pi}{4}d_z^2 SZn \tag{2.1}$$

where d_z is the diameter of the piston, S is the piston stroke, Z is the number of pistons, and n is the velocity.

For the axial piston swash-plate pump, the piston stroke changes with the swash plate angle γ as follows

$$S = D_f \tan\gamma \tag{2.2}$$

where D_f is the diameter of the piston distribution in the cylinder barrel. Therefore, the theoretical displacement of the oblique axial piston pump can be described as

$$Q_l = \frac{\pi}{4}d_z^2 D_f nZ \tan\gamma \tag{2.3}$$

Considering the leakage, the real flow rate of the axial piston pump is

$$Q_s = \frac{\pi}{4}d_z^2 D_f nZ \tan\gamma \eta_v \tag{2.4}$$

where η_v is the volumetric efficiency.

In general, the piston moves in the plunge cavity back and forth at variable speeds when the pump rotates at a constant speed. As a result, the hydraulic pump produces pulsating flow. The transient flow pulsation of a single piston can be described as

$$Q_{ti} = \frac{\pi d_z^2}{4}\omega R_f \tan\gamma \sin\alpha_i \tag{2.5}$$

where ω is the angular velocity of the cylinder barrel and α is the rotation angle of the cylinder barrel. The transient flow of the pump can be described as

$$Q_t = \sum_{i=1}^{Z_0} Q_{ti} = \frac{\pi d_z^2}{4}\omega R_f \tan\gamma \sum_{i=1}^{Z_0} \sin\left[\alpha_i + (i-1)\frac{2\pi}{Z}\right] \tag{2.6}$$

where Z_0 is the number of pistons at the discharge area (Figure 2.15) [33].

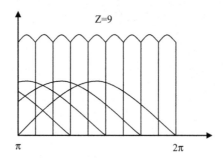

FIGURE 2.15 The flow rate ripple curve.

In a hydraulic pump, there are two types of leakage: internal leakage and external leakage. The internal leakage is proportional to the pump pressure difference:

$$Q_{il} = C_{il}P_l \qquad (2.7)$$

where C_{il} is the internal leakage coefficient and P_l is the pressure difference across the pump.

Likewise, the external leakage in each chamber is proportional to the particular chamber pressure and may be written as follows:

$$Q_{el} = C_{el}P_l \qquad (2.8)$$

where C_{el} is the external leakage coefficient and P_l is the pressure difference between the forward chamber and the return chamber.

The performance and reliability of an EDP should meet the specifications of MIL-P-19692 and AS595. The hydraulic fluid should be selected according to AS1241, and the fluid cleanliness level should meet the SAE AS 4059 class 9 standard. Figure 2.16 shows the EDP performance at flight idle speed and Figure 2.17 shows the EDP performance at takeoff speed. Figure 2.18 gives the performance at two engine speeds.

EDP testing is conducted to validate that the pump does not have excessive pressure ripple and that it can regulate pressure well for very low shaft speeds.

The high performance of an EDP is also related to improved shaft seal design, improved compensator design, higher capacity shaft bearing, hardened shoe backflange, and the replaceable valve plate.

2.3.1.2 AC Motor-Driven Pump [9,13]

In addition to the EDP, the aircraft hydraulic system also utilizes an ACMP as the primary source of supply of pressurized hydraulic fluid for the flight control and utility systems in the triplex hydraulic system structure. The same ACMP is used as a backup pump or as an auxiliary power in the main hydraulic system. The ACMP is a pressure-compensated, variable-displacement, axial piston pump driven by a constant frequency fan-cooled AC induction

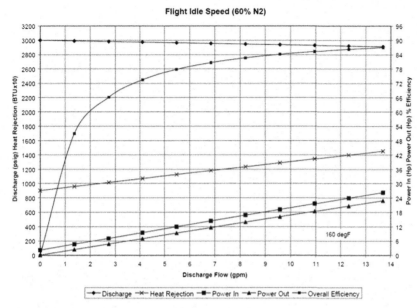

FIGURE 2.16   EDP performance at 60% N2 [9].

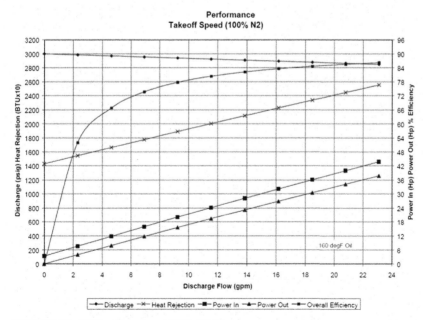

FIGURE 2.17   EDP performance at 100% N2 [9].

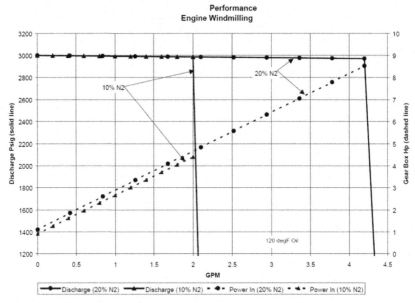

FIGURE 2.18   EDP performance at two assumed engine windmilling speeds.

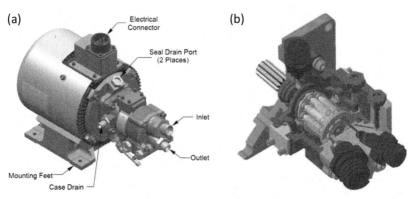

FIGURE 2.19   AC motor-driven pump. (a) External view of ACMP, (b) a cross-section of ACMP [9].

motor shown in Figure 2.19 [9]. ACMP is mounted by four mounting feet to attach to the aircraft structure. An AC electrical power supply system provides the ACMP power by an electrical connector. The discharge, suction, and case drain ports of the pump directly connect to the aircraft hydraulic system through hydraulic hoses or tubing.

The aircraft's air-cooled electric motor rotates the drive shaft and the connected cylinder block and pistons. Pumping action is generated by the reciprocating pistons, which slide on the shoe-bearing plate in the yoke assembly. Because the yoke is at an angle with respect to the drive shaft, the rotary motion of the shaft is converted to reciprocating piston motion in the rotating cylinder block.

As the drive shaft rotates the cylinder block, the piston begins to withdraw from the cylinder block. System inlet pressure is boosted by the impeller section of the motor pump and is drawn into the piston bore through a porting arrangement in the valve plate. The piston shoes are restrained in the yoke by a hold-down plate and a hold-down retainer during the intake stroke.

As the drive shaft continues to turn the cylinder block, the piston shoe continues following the yoke bearing surface. This begins to return the piston into its bore, toward the valve block. The fluid contained in the bore is compressed and then expelled through the valve block outlet port. Discharge pressure holds the piston shoe against the yoke bearing surface during the discharge stroke and provides the shoe pressure balance and fluid film through an orifice in the piston and shoe subassembly.

With each revolution of the drive shaft and cylinder block, each piston goes through the complete pumping cycle described above, completing one intake and one discharge stroke. High-pressure fluid is ported out through the valve block to the pump outlet.

Internal leakage keeps the pump housing filled with fluid for lubrication of rotating parts and for cooling. The leakage is returned to the system through a case drain port. The case relief valve protects the pump against excessive case pressure, relieving it to the pump inlet.

The principle of the ACMP is similar to the EDP. The difference is that the pump is driven by an AC electrical motor. The seal drain fitting is provided to connect the AC motor and pump. The performance of an ACMP is shown in Figure 2.20.

The characteristics of the electric motor-driven pump are as follows:

- Power on reset (POR) voltage: 115/200 VAC
- Frequency: 400 Hz
- Rated operating current: 24 A rms (max)
- Starting current: 200 A max
- Rated full flow pressure: 2700 psi (min)
- Rated full flow: 3.0 gpm
- Rated power factor: 0.8 lag (min)
- Rated operating pressure: 2000 psi (21 MPa)
- Maximum pressure: 4500 psi (32 MPa)
- Burst pressure: 7500 psi (MPa)

2.3.1.3 Hydraulic Pump Design

The aircraft hydraulic pump is a variable-displacement, pressure-compensated unit that adjusts the volume of fluid delivered to maintain constant pressure. The design of an aircraft hydraulic pump should be qualified according to

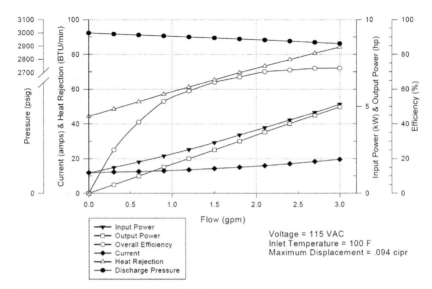

FIGURE 2.20 Performance of an ACMP [9].

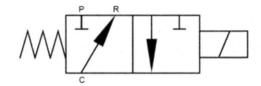

FIGURE 2.21 Pressure control three-way valve [9].

MIL-P-19692 [34] and AS595 in both mineral-based fluid and phosphate ester hydraulic fluid. A hydraulic pump can be driven by both aircraft and helicopter engines and an AC motor. A hydraulic pump should provide

- Standard rotating group, delivering constant-pressure variable-displacement fluid
- Long life under high speed and high pressure
- High-performance pump control system, which can provide an accurate response to rapid changes in flow demand
- Minimum pressure ripple to increase system reliability

In a hydraulic pump, the displaced volume is controlled by the inclined angle of the swash plate. The swash plate angle is controlled by the discharge pressure through the compensator valve and stroking piston.

A hydraulic pump utilizes the two-position, three-way solenoid valve to control the pump outlet pressure shown in Figure 2.21, in which the valve position is controlled by the discharge pressure on the valve spool area versus the preset compensator spring force. If the outside load changes, then the valve spool moves and changes the swash-plate angle.

When the solenoid valve is in the "de-energized" condition, the discharge valve is blocked, and the control port connects the case. The EDP operates as a normal pressure-compensated variable displacement piston pump. In the "energized" position, the discharge pressure is routed to the control port with return to the case port blocked at the valve. The discharge pressure is ported directly behind the auxiliary stroking piston, which pushes on the primary stroking piston, decreasing the swash plate and reducing the pump outlet flow (Figure 2.22). When the command to the solenoid is removed, the valve returns to the closed position, venting the pressure applied behind the stroking piston back to the case.

Figure 2.22 shows the stroking piston nearly fully extended, putting the swash plate at a very slight angle and generating small output flow. The high-pressure discharge fluid connects to a compensator valve spool. If the user requires a high flow rate, then the discharge pressure decreases and the spring forces the spool to move to the right position. When the fluid pressure reaches the value defined by the setting of the spring, the spool valve transfers the fluid from the stroking piston to the case pressure. Increasing the swash plate angle will increase the output flow rate. The compensator valve is the integration from the volume of fluid to the stroking piston and pump output. If the flow demand drops and the discharge pressure rises above the compensated value, then the valve passes fluid to extend the stroking piston and the swash plate angle decreases to lower the discharge pressure.

2.3.2 Power Transfer Unit [9,18,19]

The PTU is a fixed-displacement, unidirectional device used to transfer fluid power from hydraulic system 1 to hydraulic system 2 without sharing fluid between systems. The PTU includes two fixed-displacement axial piston rotating groups housed separately and connected through a common mounting plate by a driveshaft, shown in Figure 2.23. Because the PTU can only operate in one direction, the motor side is somewhat larger in displacement than the pump side to maintain system pressure. The case drain port of the pump side is used to route internal leakage, cooling, and idle bypass flow back to the reservoir. The idle bypass flow permits the smooth and continuous operation of the PTU at all demands.

A fixed-displacement, unidirectional PTU consists of a hydraulic motor and a hydraulic pump typically bolted together through a common mounting plate and coupled by a drive shaft, as shown in Figure 2.24.

The PTU characteristics are similar to the EDP and ACMP. The three-way valve controls the pump outlet pressure, as shown in Figure 2.25, in which the valve position is the result of the interaction between the discharge pressure on the valve spool area versus the preset compensator spring force. The outlet

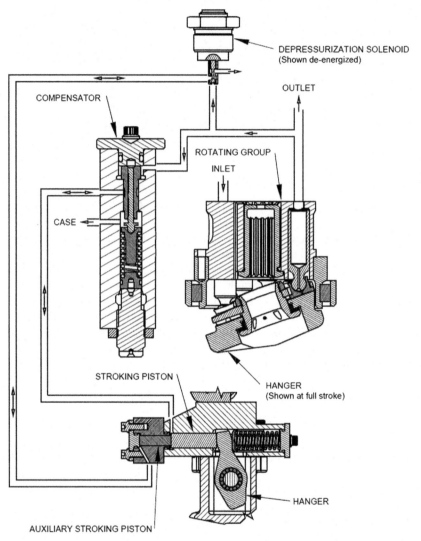

FIGURE 2.22 EDP pressure control circuit.

pressure change will lead to the hanger angle changes. The hanger angle is controlled by the position of the stroking piston. When the stroking piston is nearly fully extended, the hanger angle decreases to a very small value and generates small output flow. If the high flow rate requirement makes the discharge pressure decrease, then the spool moves to the right because of the spring force. When the pressure reaches the value defined by the setting of the spring, the spool uncovers a port in the sleeve and the valve transfers fluid from the stroking piston to the case pressure.

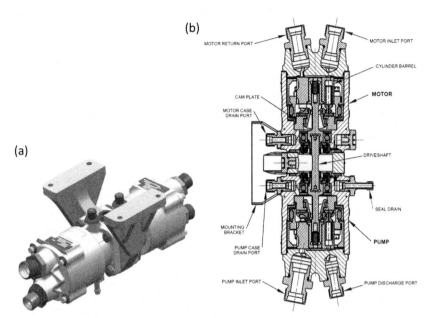

FIGURE 2.23   The structure of a PTU. (a) PTU outside drawing, (b) PTU cross-sectional drawing.

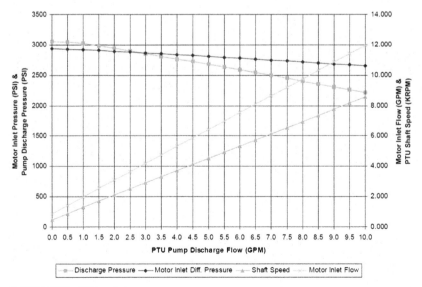

FIGURE 2.24   PTU performance.

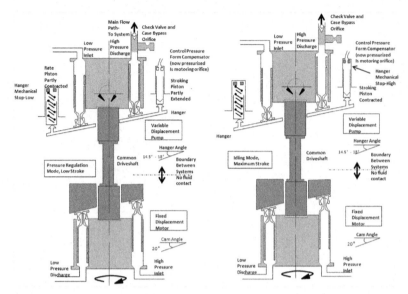

FIGURE 2.25   PTU control schematic.

The piston extends to drive the pump hanger to a larger inclined angle as the stroking piston retracts. The higher cam angle increases the flow output. If the flow demand decreases and the discharge pressure increases, then the valve passes the fluid to extend the stroking piston and the hanger decreases the cam angle to lower the discharge pressure.

In aircraft with a green hydraulic system and yellow hydraulic system, the PTU enables the green system to be supplied by the yellow system and vice versa without fluid exchange between the two systems. The PTU ensures essential flight functions by supplying the green system with additional power at constant pressure for the gear retraction should engine 1 fail during takeoff, and similarly for the yellow system elevator servo control in the event of a dual failure of the blue and the yellow hydraulic systems based on a failure of engine 2.

Ground maintenance functions are also facilitated by the PTU while the engines are shut down. The subsystems associated with the green system can be operated on the ground by transferring power from the yellow system. Power to the yellow system is supplied by the onboard electric motor pump or possibly through a ground service power supply. In addition, the bidirectional capability of the unit makes possible the operation of the yellow system by means of transferring power by a ground service cart attached to the green system.

The only external control function of the PTU is accomplished via a switch in the cockpit, which energizes valves in the PTU supply lines of the yellow and green systems. This permits the complete isolation of the PTU and prevents hydraulic power transfer to or from either system. These valves can be energized by the crew in flight at any time, especially during warning indications, or they can be activated by inhibition logic when the aircraft is on the ground during engine startup.

2.3.3 Priority Valves [9,20]

The function of a priority valve is to allow hydraulic fluid flow to certain functions within a hydraulic system when the pressure is greater than or equal to a specified level. In effect, the priority valve gives priority to certain components over less critical components through cutting off hydraulic power to heavy load users. For example, in a 3000-psi (21 MPa) hydraulic system, a 2000-psi (14 MPa) priority valve may be installed such that components downstream of the priority valve are only supplied hydraulic pressure when the pressure in the system is above 1500 psi (10.5 MPa). Below 1500 psi (10.5 MPa), the priority valve will be closed.

A typical application of a priority valve is to preserve hydraulic pressure for primary flight controls. When pressure drops below a specified level, one or more priority valves will close, leaving flow available to primary flight control HAs. For example, in F-14, when the EDP pressure drops below 2400 psi (17 MPa), a priority valve isolates the landing gear, brakes, nose wheel steering, ram air door servo actuator, over-wing fairing, and other components. When the isolation valve shuts at 2400 psi (17 MPa), the EDPs provide hydraulic pressure to the pitch servo, roll servo, yaw servo, rudders, speed-brakes, and hook lift. In other applications, a hydraulic system is powered by the main system pump and an auxiliary pump. Under normal operation of the main system pump, all hydraulic services can be powered. However, should the main pump fail, the auxiliary pump would take over. The auxiliary pump may be powered through a PTU or an electric motor. Auxiliary pumps are normally sized smaller than the main system pump because under normal operation there is a weight penalty when the system is not being used. If the auxiliary pump has a lower output pressure, then a priority valve can be used to isolate noncritical components at the lower auxiliary pump pressure capability.

The priority valve consists of a poppet and spring, as shown in Figure 2.26. When inlet pressure P1 is sufficient to overcome downstream pressure P2, valve friction, and spring force, then fluid flows through the valve; otherwise, the valve will be closed. By setting the spring force sufficiently high so that

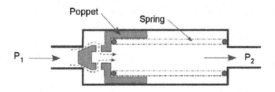

FIGURE 2.26 Structure of priority valves.

high pressure is required to overcome the spring force, the valve can act like a priority valve. Some priority valves allow free flow in the reverse flow direction (the priority valve shown in Figure 2.26 does not allow reverse flow).

2.3.4 Control Valves [17]

Control valves are valves used to control conditions such as flow, pressure, temperature, and liquid level by fully or partially opening or closing in response to signals received from controllers. The most common final control element in the control system is the control valve. The control valve manipulates hydraulic fluid to compensate for the load and maintain the process variable as close as possible to the desired set point. The control system usually consists of a sensor, transmitter, and controller that compares the "process variable" received from the transmitter with the "set point." The controller sends the corrective signal to the control valve to control the condition. It is obvious that the control valve provides the interface between the hydraulic power elements (i.e., the pump and the hydraulic output device). Within valve-controlled hydraulic circuits, the control valve receives feedback from the operator and adjusts the system output accordingly.

Hydraulic control valves are classified in several ways. The most general classification is based on the number of flow lines connected to the valve. For example, a three-way valve has three flow lines given by a supply line, an output line, and a return line back to the reservoir. Hydraulic control valves may also be classified either by construction or by function. Construction-based valves include spool valves, poppet valves, etc. Function-based valves consist of directional-control valves, flow-control valves, and pressure-control valves. The directional-control valves are used as switching devices within the hydraulic circuit. Flow-controlled valves are used to modulate and continuously direct flow within the hydraulic system. These valves perform the control objective by implementing some form from the output device. If the feedback is generated automatically through the internal sensing and control mechanism, then the valve is called a *servo valve*. Pressure-controlled valves are used to maintain or limit a specific pressure level in a hydraulic circuit.

Hydraulic control valves are widely used in hydraulic control systems. The sliding valves are the most commonly used means in an aircraft hydraulic system such as directional control valve, spool valve, and piston valve. Figure 2.27 shows the high-performance, direct-operated, four-way valves made by Parker. They are available in two- or three-position styles. These valves were designed for industrial and mobile hydraulic applications that require high cycle rates, long life, and high efficiency.

The directional control valves consist of a four-chamber style body and a case-hardened sliding spool. The spool is directly shifted by various operators, including a solenoid, lever, cam, air, or oil pilots. Figure 2.28 shows its flow versus pressure drop curve.

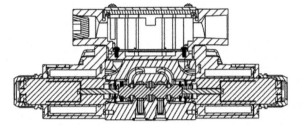

FIGURE 2.27   Structure of solenoid operated plug-in conduit box style.

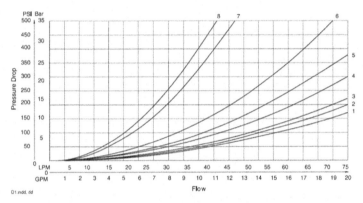

FIGURE 2.28   The flow versus pressure drop of a directional valve.

2.3.5 Check Valve [1,21]

Check valves, Figure 2.29, are two-port valves—one for fluid to enter and the other for fluid to leave. Normally, the check valve closes under the action of the spring force on the left. When the fluid flows from right to left and the fluid pressure is greater than the cracking pressures, the spool moves off of its seat and opens the valve. An important concept in check valves is the cracking pressure, which is the minimum upstream pressure at which the valve will operate. Typically, the check valve is designed for controlling of fluid flow in one direction; therefore, it can be specified for a specific cracking pressure.

The check valves are leak-free in one direction and free-flow in the reverse direction. The fluid flow direction can be controlled with a check valve. The pilot-operated check valve is another type of check valve, Figure 2.30.

A pilot-operated check valve can be opened by a pilot signal. High-pressure fluid enters the pilot port and opens the pilot valve and check valve against the load pressure and spring force. In general, check valves are very small, simple, and/or inexpensive. Check valves work automatically, and most are not controlled by a person or any external control; accordingly, most do not have any valve handle or stem.

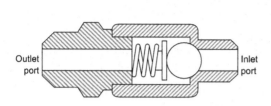
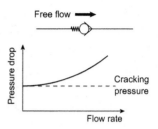

FIGURE 2.29   Check valve structure.

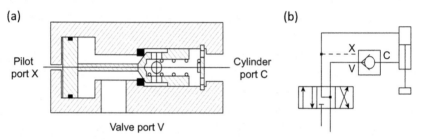

FIGURE 2.30   Pilot-operated check valve.

2.3.6 Hydraulic Accumulator [1,22]

The hydraulic accumulator, Figure 2.31, is an energy storage device in which one end is closed and another is connected to the hydraulic pipes. The hydraulic accumulator is divided into three parts: compressed gas (air chamber), piston, and hydraulic fluid (oil chamber).

Hydraulic fluid, pressurized by a hydraulic pump, forces the piston of the accumulator to compress the gas in the air chamber. The compressed gas can store the energy just like a spring. When the aircraft needs the extra flow, the compression energy will release to compensate for the system needs. The main function of the hydraulic accumulator includes the following:

- Provide transient flow in a short time: The hydraulic accumulator can provide the peak flow in a short time in an intermittent work hydraulic system such as the landing gear control and flap control. The hydraulic accumulator can decrease the size and reduce the weight of the hydraulic system.
- Compensate for the leakage and maintain the pressure: The accumulator can provide the flow and maintain constant pressure for the actuation unit that does not move for a long time but needs to maintain constant pressure.
- Work as an emergency power supply: When the hydraulic power supply system suddenly stops feeding the hydraulic fluid, the hydraulic

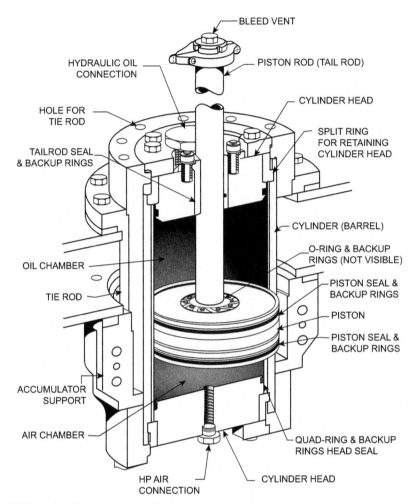

FIGURE 2.31 Hydraulic accumulator structure.

accumulator can provide the fluid to keep the system operational in case of emergency.

- Absorb the pressure ripple of the hydraulic system: Because the hydraulic system has inherent flow ripple and fluid solid coupling vibration, its pressure ripple will influence the actuation system performance. The accumulator can absorb the pressure ripple and maintain the pulsation within the allowable range.
- Absorb the impact pressure: When the control valve suddenly changes the direction or the actuation system suddenly stops, the accumulator can absorb their impact pressure.

2.3.7 Hydraulic Filter [1,24]

Because some hydraulic components, such as the SV, are sensitive to debris in fluid, it is necessary to keep the hydraulic fluid clean. The hydraulic filter is used to eliminate contamination within the hydraulic fluid, in which the filter element could stop all particles of contaminant above 5 μm in size and a high percentage of particles below this size. The hydraulic filter consists of three parts: head, filter element, and bowl, as shown in Figure 2.32.

The head of the hydraulic filter is secured to the hydraulic pipes. Considering the filter block, some hydraulic filter heads have a pressure-operated bypass valve, which can route the hydraulic fluid directly from the inlet to the outlet port in the case of filter blockage. The bowl of the filter is used to hold the element to the filter head. The filter element is of the 5-μm noncleanable, woven mesh, porous metal, or magnetic type. When the particles, 5 μm or larger, go through the filter, the filter element stops the particles and allows the clean fluid through. The hydraulic filter can control the particles in fluid and maintain the hydraulic system at some cleaner levels. The most common 5-μm filter medium is composed of organic and inorganic fibers integrally bonded by epoxy resin and faced with a metallic mesh upstream and downstream for protection and added mechanical strength. The 5-μm, noncleanable, hydraulic filter elements should be replaced with new elements during specified maintenance inspection intervals in accordance with the applicable procedures. Another type of 5-μm filter medium is fine stainless steel fibers, which draw into the particles and sinter the

FIGURE 2.32   Hydraulic filter assembly.

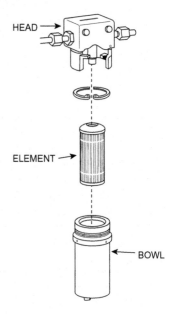

material into a thin layer with controlled filtration characteristics. These types of filter elements may be cleanable or noncleanable, depending upon their construction, and are marked accordingly.

The hydraulic filter is always located in the suction line, pressure circuit, return line, and branch circuit.

2.3.8 Hydraulic Reservoir [1,25]

The hydraulic reservoir is used to hold the volume of fluid, transfer heat from the system, allow solid contaminants to settle, and facilitate the release of air and moisture from the fluid. The reservoir, Figure 2.33, has its own housing and is connected with other hydraulic components by a tubing or a hose. The space above the normal level of the fluid is used for fluid expansion and for escape of trapped air.

FIGURE 2.33   Hydraulic reservoir of an aircraft.

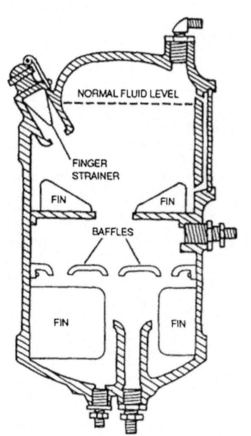

There are two types of hydraulic reservoirs: vented reservoirs and pressured reservoirs. In vented reservoirs, atmospheric pressure and gravity are the forces that cause fluid to flow out of the reservoir and into the pump. To supply a pump with fluid, a vented reservoir must be positioned at a higher location than the pump. If the reservoir and the pump were at the same level, then gravity would have no effect on fluid flow. If the reservoir was at a level below the pump, then fluid would tend to run out of the pump and into the reservoir. In this case, the pressurized reservoir can force the fluid upward into the pump. If the aircraft flies at very high altitudes, then atmospheric pressure decreases as the altitude increases. In this case, it is also necessary to use the pressurized reservoirs to force fluid into the pump.

The fluid viscosity is related to the fluid temperature. The higher the temperature, the less viscous the fluid. If the fluid with low viscosity cannot provide normal pressure for the pump, then the flight control system cannot complete its functions. Therefore, it is necessary to design the heat exchange system for the hydraulic power supply system.

Figure 2.34 shows the aircraft layout of hydraulic reservoirs with a System A reservoir, System B reservoir, and standby reservoir.

2.3.9 Fluid Cooling System [1]

As the hydraulic actuation system drives the control surfaces, it generates heat, thus requiring a cooling system to remove the heat. The main component of the cooling system is the heat exchanger. There are two types of heat exchangers: aluminum plate-fin heat exchangers and aluminum flat-tube heat exchangers. Aluminum plate-fin heat exchangers, Figure 2.35, consist of finne passages separated by flat plates with a unique internal configuration to maximize heat transfer. Aluminum flat-tube heat exchangers consist of several flat tubes with multiple extended surface channels within each tube. Fins are vacuum-brazed between the flat tubes and form the passages for the second fluid. Heat exchangers are low-pressure devices that are normally situated in the return line of the actuator.

FIGURE 2.34   Aircraft layout of hydraulic reservoirs.

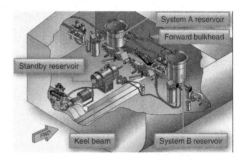

FIGURE 2.35   Plate-fin oil-cooling heat exchanger.

2.3.10 Hydraulic Actuator [1,12,23]

The actuator is a mechanism that is responsible for delivering force and motion to the control surfaces. There are two types of actuators in real application: linear actuators and rotary actuators. Linear actuators are also called hydraulic cylinders, which generate linear output for the hydraulic control system. The rotary actuator is also called the hydraulic motor, which generates rotational output for the hydraulic system. Here we use the linear actuator as an example to illustrate its design principle.

The conventional linear actuator, Figure 2.36, is the hydraulic SV-controlled cylinder, in which the SV transfers the electrical signal to mechanical movement of the spool and then controls the fluid flow to drive the cylinder movement.

2.3.10.1 Technical Requirements

Before designing the actuator, it is necessary to know the technical requirements for an actuator, as follows:

- Maximum displacement: X_{max}
- Maximum speed: V_{max}
- Maximum force: F_{max}
- Rated hydraulic system pressure: P_S
- Breakaway pressure: P_b

2.3.10.2 Static Design of Actuator

2.3.10.2.1 The Determination of Cylinder Size

According to the load-matching regulation, the best design is to satisfy the maximum load at the maximum power point. Because the load pressure at the maximum power point is $P_L = 2/3P_S$, the piston area of the actuator can be calculated as

$$A = \frac{F_{max}}{P_L} = \frac{F_{max}}{\frac{2}{3}P_S} \tag{2.9}$$

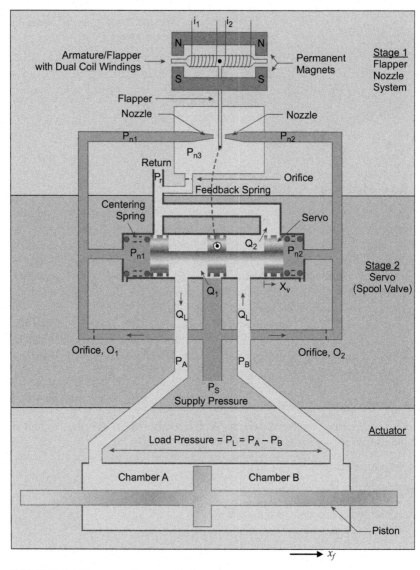

FIGURE 2.36 Flapper nozzle servo actuator.

If the single-rod hydraulic cylinder is selected, then the piston diameter D and the piston rod diameter d can be calculated by the following equation:

$$A = \frac{\pi(D^2 - d^2)}{4}$$

(2.10)

2.3.10.2.2 Servo Valve Determination

The SV should meet the requirement of actuator flow and achieve the dynamic performance of an actuator. Because the frequency of the actuator does not exceed 10 Hz and the maximum motion speed is V_{max}, the load flow rate of the actuator can be obtained at the maximum power point as follows:

$$q_0 = \sqrt{3} A V_{max} \qquad (2.11)$$

Considering the leakage, the flow rate of the SV can be selected as:

$$q_{sv} > q_0 \qquad (2.12)$$

2.3.10.2.3 Hydraulic Power Supply System Flow Rate Determination

If the actuation system consists of n actuators, and each actuator needs flow rate q_{sv}, then the hydraulic power supply system flow rate can be determined considering the leakage as

$$Q_l = n\left(q_{sv} + \Delta\right) \qquad (2.13)$$

where Δ is the leakage flow rate.

2.3.10.3 Dynamic Design

In the dynamic design of a linear actuator, the mathematical model is used to represent actuation-system dynamics. HA consists of electrical amplifier, SV, cylinder, and displacement sensor (a linear variable differential transformer (LVDT)). The diagram of a linear HA is shown in Figure 2.37.

2.3.10.3.1 Electrical Amplifier Modeling

According to the torque motor voltage equation, the electrical amplifier model can be described as

$$i = K_u \Delta u \qquad (2.14)$$

where K_u is the gain of the electrical amplifier and i is the input current to the SV.

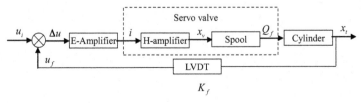

FIGURE 2.37   Block diagram of a linear actuator.

FIGURE 2.38   Block diagram of a cylinder.

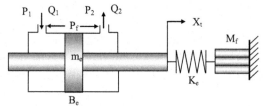

2.3.10.3.2 SV Modeling

As the frequency width is more than 100 Hz whereas the frequency width of the linear actuator is 10 Hz, the hydraulic SV can be expressed by the first inertial model:

$$Q_f = \frac{K_{sv}}{T_{sv}s + 1}i \tag{2.15}$$

where K_{sv} is the SV gain, T_{sv} is the time constant of the SV, and Q_f is the flow rate of the SV.

2.3.10.3.3 Cylinder Modeling

The load of the HA can be considered as the inertial load and flexible load shown in Figure 2.38.

The spool flow rate equation is

$$Q_f = K_q x_v - K_c P_f \tag{2.16}$$

where Q_f is the load flow (m³/s), K_q is the flow gain of the SV (m³/s/A), x_v is the spool displacement of the SV (m), K_c is the pressure-flow rate coefficient of the SV (m³/s/Pa), and P_f is the load pressure.

The continuous flow equation of the cylinder is

$$Q_f = A\frac{dX_t}{dt} + C_{sl}P_f + \frac{V_t}{4E_y}\cdot\frac{dP_f}{dt} \tag{2.17}$$

where A is the piston area (m²), X_t is the piston displacement (m), C_{sl} is the leakage coefficient of the cylinder (m³/s/Pa), V_t is the volume of the cylinder (m³), and E_y is the equivalent elasticity modulus (N/m²).

The force balance equation is

$$AP_f = m_f\frac{d^2X_t}{dt^2} + B_e\frac{dX_t}{dt} + K_e X_t \tag{2.18}$$

where m_f is the equivalent mass (kg), B_e is the viscous damping coefficient of the cylinder (Nm/s).

The feedback equation is

$$u_f = K_f x_t \tag{2.19}$$

where K_f is the feedback gain.

Then, the transfer function of the SV-controlled cylinder can be described as

$$\frac{X_f}{Q} = \frac{\frac{1}{A}}{s\left(\frac{s^2}{\omega_h^2} + \frac{2\zeta_h}{\omega_h} + 1\right)} \tag{2.20}$$

where ζ_h is the hydraulic damping coefficient and ω_h is the natural frequency of the hydraulic cylinder. The open-loop transfer function of the linear actuator can be described as

$$G(s) = \frac{K_V}{s(T_v s + 1)\left(\frac{s^2}{\omega_h^2} + \frac{2\zeta_h}{\omega_h} s + 1\right)} \tag{2.21}$$

The block diagram of the linear actuator is shown in Figure 2.39.

The corresponding Bode plot of an open-loop linear actuator is shown in Figure 2.40.

A closed-loop Bode diagram is shown in Figure 2.41 when the proportional-plus-integral-plus-derivative (PID) compensator is added in the actuator.

The above model is a linear model, which is used for simulation analyses, to evaluate the response to step (0.03 m at 0.5 s) and sinusoidal input (0.03 m, 1 Hz) demands for a linear actuator shown in Figures 2.42 and 2.43.

Figures 2.42 and 2.43 show that the response of the linear actuator can well follow the command.

2.3.10.4 Nonlinear Parameter Influence

The mathematical model of an actuator can indicate the frequency-response characteristics (the response amplitude and phase lag when responding to a sinusoidal input demand). Linear models can be created and the frequency-response characteristics determined using traditional linear analysis techniques. In this case, the actuator design needs to meet specified frequency-response characteristics. This is particularly important in flight control system design because the aircraft control laws are designed with an assumed gain and phase characteristic for the actuators (usually in the form of a second- or third-order transfer function), and any significant deviation from these characteristics can lead to a reduction of aircraft gain and phase stability margins. However, the linear

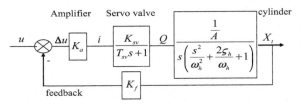

FIGURE 2.39   The block diagram of a linear actuator.

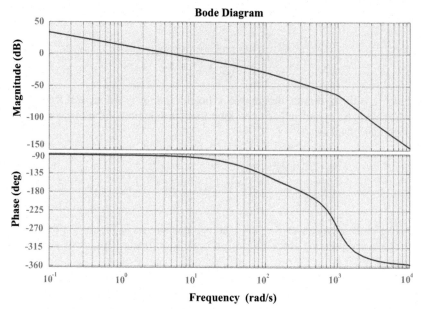

FIGURE 2.40   The Bode plot of a linear actuator.

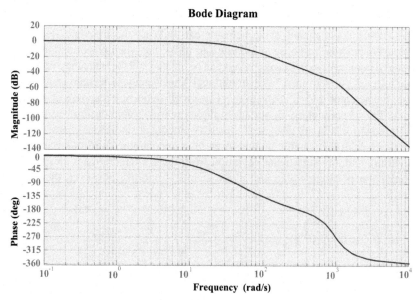

FIGURE 2.41   The Bode plot of a closed-loop actuator.

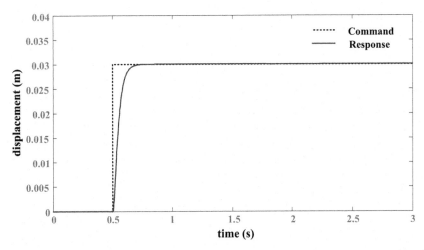

FIGURE 2.42   Step response under 0.3 m at 0.5 s.

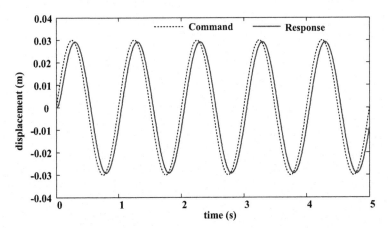

FIGURE 2.43   Sinusoidal input response under 0.3 m and 1 Hz.

model only shows an actuator frequency response under moderate amplitude conditions, but in reality the response would vary with demand amplitude.

There are several nonlinear parameters influencing the dynamic performance of HAs, such as the flow-pressure nonlinear characteristic of the SV, saturated characteristics, and friction. For example, the flow-pressure relationship can be described as

$$Q_f = C_v W x_v \sqrt{\frac{p_s - p_f}{\rho}} \qquad (2.22)$$

where C_v is the flow gain at the valve window and W is the area gradient of a spool.

If the sinusoidal demands inject the nonlinear model of the HA, then the gain and phase relationship between demand signals and the resulting actuator response under different amplitudes is shown in Figure 2.44. Larger demand amplitudes of the actuator will reach internal limits, such as current or voltage limits in motors and spool displacement limits in SVs and main control valves. The relationship between the limits reached in terms of amplitude and the frequency of the demand signal have a significant effect on the nonlinear frequency response of the actuator and can lead to stability problems.

Considering the SV position limit, the response of the ram will be affected with a consequent effect on frequency-response characteristics when these limits are reached, as shown in Figure 2.44 [35]. Figure 2.45 shows that the gain reduces and the phase lag increases as the demand amplitude increases using a nonlinear model based on the block diagram [35]. In this case, a demand amplitude of 0.5 mm represents a linear response because neither the spool nor the main control valve position limits are reached. Even at demand amplitudes of up to 2 mm across the considered frequency range considered (up to 40 Hz), the internal limits are not reached; therefore, the actuator responds in a linear fashion. However, at higher amplitudes, the limits are reached and the frequency response is affected. However, it should be noted

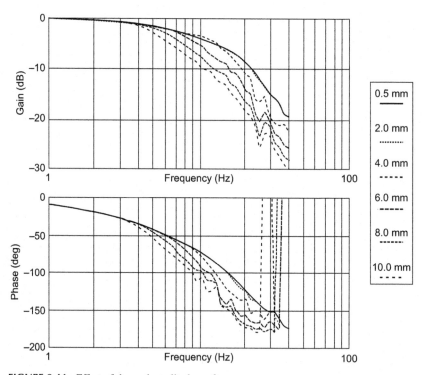

FIGURE 2.44   Effect of demand amplitude on frequency response.

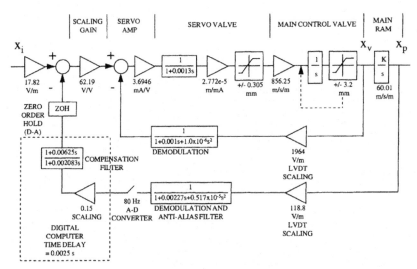

FIGURE 2.45   Block diagram of actuation system with nonlinear factors.

that the demand amplitudes to reach saturation are quite high for the demand frequency at which the saturation occurs. For example, at a demand of 4 mm, the limits are not reached until approximately 10 Hz, and at 6 mm the demand signal must be at 6 Hz for the saturation to occur. The actuator would not be expected to operate at these combinations of demand amplitude and frequency and can be proven to meet the specified frequency-response characteristics for all valid operating conditions. In practice, the actuator would be designed for higher rates than necessary for aircraft control and stability to provide some growth potential. Limits on position demand and rate would be applied in the control law demands on the actuator.

2.3.10.5 Saturation Analysis [35]

Saturation analysis is a technique closely related with the production of nonlinear frequency-response data. A linear model of the actuator is used to calculate the frequency response from input-demand signals to the locations of the various limits, such as SV position or main control valve position. The gain information from this analysis can then be used to determine the demand amplitude, which will cause the limit to be reached across a range of frequencies. Figure 2.46 shows the results of such an analysis on the basis of the actuator model. It can be seen that saturation will not occur for sinusoidal demands of up to 2 mm in amplitude. For demands between 2 and 4 mm, the SV position limit will be reached if the frequency of the demand signal is above 12 Hz, but the main valve limit will not be reached for any demand signal. For demands with a frequency below 12 Hz, the main control valve position limit will be reached before the SV position limit if the amplitude of the demand signal is higher than 4 mm. A demand signal of 6-mm amplitude

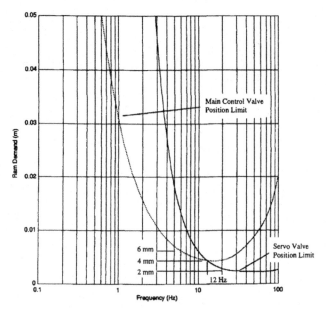

FIGURE 2.46   Saturation analysis results [35].

will cause the main control valve limit to be reached if the frequency is above 6 Hz.

From the nonlinear frequency response discussed above, these results indicate that there are no nonlinear effects likely to affect the actuator response in the standard condition. However, it is possible to have combinations of gains and limits around the control loop that can cause a jump resonance effect, with serious performance implications for the actuator, as discussed below.

2.3.10.6 Jump Resonance [35]

Under conditions of large-amplitude demand when SV travel limits are encountered, an actuation system can display sudden large increases in phase lag. This phenomenon is described by the term *jump resonance* and is caused by an effective acceleration limit.

In practice, if in some extreme maneuvers it is possible to reach such limits, then the additional phase lag caused by the jump resonance can lead to a severe temporary reduction of aircraft stability margins, with consequent potential handling difficulties. To avoid this problem at the design stage, it is important to ensure that the valve ports (for the main control valve and SV) and travel limits are adequately sized. Increased valve port width can be compensated for by a reduction of electrical gain, maintaining the overall loop gain required for actuator stability and response.

Jump resonance is characterized by a very rapid increase in phase lag over a narrow frequency band, as shown in Figure 2.47 [35]. This figure shows

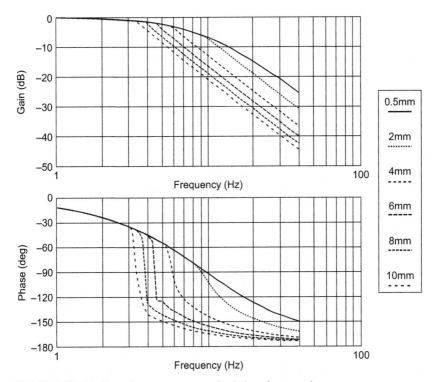

FIGURE 2.47 Nonlinear frequency-response simulation of actuator jump resonance.

nonlinear frequency response results obtained from a model configured to exhibit jump-resonance effects. With the increase of demand amplitude, the frequency at which saturation occurs reduces and a reduction in gain is seen, along with an even more dramatic increase in phase lag. The potential jump resonance can be predicted from saturation analysis results. Figure 2.48 shows the saturation characteristics for the actuator with jump resonance [35].

Saturation of the SV position limit will occur before saturation of the main control valve position, leading to an effective actuator acceleration limit for most demand amplitude and frequency combinations. The crossover frequency is at a relatively low frequency (2.5 Hz), which is within the bandwidth of aircraft control. An actuator such as this is likely to cause severe handling deficiencies in an aircraft, and the valve ports and control gains would need to be redesigned to give a better balance.

2.3.10.7 Failure Transients [35]

Although redundancy features are included in the actuator design to ensure continued operation after failures, it is also important that the actuator does not produce an excessive transient when transitioning from one level of

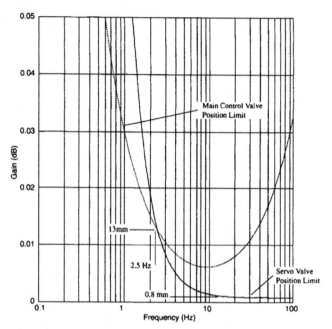

FIGURE 2.48   Saturation simulation of actuator with jump resonance [35].

redundancy to another. Boundaries, within which the transient must lie, are defined in an actuator specification. Failure to comply with this requirement will lead to excessive actuator transient movement immediately after a failure, which could produce structural damage in the area of the actuator mountings or a high transient acceleration at the crew's stations or at sensitive equipment.

The effect of the failure can be countered either by a failure-absorption method or by failure rejection. In the case of a failure-absorption method, the presence of the failure is countered by the rest of the system with no special action being taken. In the failure-rejection method, the failure is detected and an appropriate action removes the effects of the failure, leaving the remaining parts of the system to continue operation. For any methods, the transient response induced after a failure must be assessed and a design should be produced that will meet the specified requirements. For the failure-rejection approach, the failure detection and isolation algorithms must also be designed.

Failure-detection algorithms are often referred to as *built-in-test* (BIT). Several levels of BIT will exist on an aircraft actuation system, ranging from start-up checks and preflight checks automatically performed by the flight control system through to the continuous monitoring of actuator operation, referred to as *continuous built-in test* (CBIT). The level and method of CBIT depend largely on the actuator design, but the following typical example will illustrate the principles.

Ram position measurement for feedback control is often performed using LVDTs. Typically, three or four LVDTs are used to provide the necessary levels of redundancy. The individual LVDT signals are consolidated in each computer to produce an average signal that is used for actuator control-loop closure. In this way, minor build tolerances, temperature effects, etc., on each LVDT can be averaged out, minimizing the difference in drive signals between lanes, which would produce a force fighting on mechanical components. However, this method would allow fault in one of the LVDTs to be propagated to the consolidated ram position signal in all of the flight control computers. This effect of extreme signals and failed sensors can be reduced by application of averaging algorithms, which are weighted toward the median. A typical voter algorithm for a triplex system is shown in Figure 2.49.

The incoming signal samples are first sorted to determine the highest, median, and lowest values. The voter algorithm then produces a consolidated or averaged value based on the three input values. Several alternative algorithms could be used, including simple averaging of the three values or selection of the median. To minimize the influence of faults on the consolidated value, the algorithm will limit the authority of the highest or lowest signals, weighting the average toward the median.

Having produced algorithms that limit the influence of faulty sensors on control-loop closure, it is also necessary to determine which lane is faulty and to reconfigure the voter monitor to ignore that lane. This is the purpose of the BIT algorithms. In the case of the LVDT monitor under consideration here, the faulty lane could be identified by comparing each lane's signal with the consolidated value. Because we have already established that the influence of the faulty lane on the consolidated signal is limited, any lane that shows a significant difference from the consolidated value (more than a certain tolerance value) can be considered faulty. To minimize the number of nuisance failure warnings, the tolerance value is selected to allow for a difference in

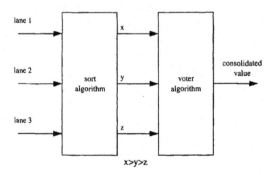

FIGURE 2.49   Triplex signal-consolidation algorithm [35].

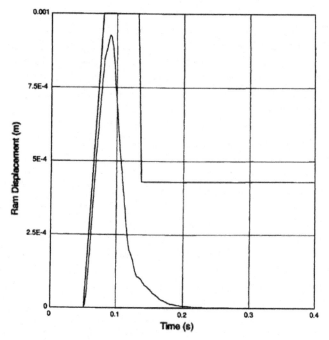

FIGURE 2.50   Transient response under LVDT failure [35].

build standard of the LVDTs and to remove the effects of false fault. In addition, a fault must be present for a certain period of time before fault determination, such as five consecutive computer iterations, before it is confirmed as a failure and the appropriate action taken. In this case, the appropriate action would be to change the voter algorithm to simple duplex averaging using only the two healthy signals.

Simulation is used in defining the voter algorithms to minimize the influence of the faulty signals and to design CBIT algorithms. An actuator model is produced, including various redundancy features and signal-consolidation algorithms. Typical faults can then be simulated and the effects on actuator response predicted, as shown in Figure 2.50. This figure represents transient response for the actuator model shown in Figure 2.45 as the outer-loop (ram-position) consolidation changes from triplex to duplex.

2.3.11 Redundant HAs [26,27,28,29,31]

To improve the reliability of HAs, a redundant technique is widely used in actuator design, especially in aircraft with a control configuration vehicle and FBW. Because the reliability of electrical component cannot increase any more, only the redundancy design could increase the system reliability to meet the growing reliability requirement. The HA is the key component of FBW, the failure of which directly affects flight performance; therefore, it is necessary to design the redundant actuator to satisfy the requirement of the aircraft flight control system. There are three key problems in redundant actuator design:

1. Force coupling: A redundant actuator is a nonlinear system with multiple inputs and coupling outputs. With the asymmetry among different channels, such as different SVs, different installation conditions, and the outside disturbance, the redundant actuator results in coupling in the output rod. The following methods can decrease the force coupling:
   a. Improve the machining accuracy and reduce the difference among channels.
   b. Decrease the system stiffness and increase the system damping.
   c. Adopt a balance technique to make each channel pressure the same.
   d. Use decoupling techniques to eliminate the cross-disturbance.
2. Redundancy management: Redundancy management is a process that can be automatically performed in full. The minimum number of monitors should be a used to decrease the complexity and alarm rate. The function of redundancy management is to maximize the use of resources and, as much as possible, to restore the existing fault component.
3. Fault monitoring: Fault monitoring is used to dynamically detect, diagnose, and isolate the fault. The conventional flight control system requires FO/FO/FS (failure-operation-failure-operation-failure-safe) redundancy design therefore, more than triplex redundancy should be used in actuator design. Different redundant actuators utilize different fault monitoring algorithms. For example, triplex redundancy always uses comparison monitoring, double redundancy adopts model-comparison monitoring, and one channel connects the two cavities of the cylinder.

From the design requirement of the flight control system, the redundant actuator should meet the following conditions:

- Satisfy the performance requirement of FBW and keep the performance similar to a single actuator.
- Satisfy the reliability requirement. Normally, commercial aircraft require that the failure rate of a redundant actuator is less than 10^{-7} per hour; therefore, it is necessary to adopt the triplex or quadruplex redundant actuator.

The detailed requirements of a redundant actuator include the following:

1. Maximum output force (torque)
2. Mechanical motion range (displacement/angle, velocity/angular velocity)
3. Actuator stiffness
4. Static performance: static transmission ratio, static output characteristics
5. Dynamic performance: response time or frequency width
6. Fault detection coverage
7. Failure operation level
8. Fault isolation mode
9. Failure rate
10. Mean time between failures (MTBF)

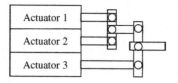

FIGURE 2.51   Displacement synthesized redundant actuator.

In redundant actuator design, the centralized redundant actuator and distributed redundant actuator are widely used in aircraft. The centralized redundant actuator has the servo controller located close to the flight control computer while the servo control cylinder is close to the control surface. Design needs space to install the pipes between the servo controller and the cylinder. The distributed redundant actuator is the one where servo controller is separate from the flight control computer and has the individual actuator close to the control surface. There are three kinds of redundant actuators, and they are

1. Displacement synthesized redundant actuator: This actuator connects the individual cylinder output with a rod. For the triplex redundant actuator shown in Figure 2.51, the swing arm is used to connect the output of cylinders. The output of this kind of redundant actuator is the mean value of the three-cylinder output. If the failure occurs, then the redundant actuator cannot compensate; therefore, its performance will degrade. Because of the complex structure, it is currently rare to use a displacement synthesized redundant actuator.
2. Electromagnetic synthesized redundant actuator: A rare earth permanent magnetic motor is a kind of high-power motor that can directly drive the control valve to construct the redundant actuator. With a rare earth permanent magnet motor, four sets of a single actuator are not required, whereas only four control coil force motors are set in a quadruply redundant actuator. Likewise, another redundant actuator can be designed with four control coil force motors and two hydraulic cylinders, as shown in Figure 2.52. This kind of design can simplify the structure of the redundant actuator.
3. Force synthesized redundant actuator: Force synthesized redundant actuators include mechanical force synthesized redundant actuators and hydraulic force synthesized redundant actuators. The common characteristic of force synthesized redundant actuators, Figure 2.53, is that the redundant cylinder is installed in a parallel layout and force synthesizing arm connection. This structure can prevent the fault transient to the synthesizing ram.

In a redundant actuation system, the following factors should be considered:

1. Load sharing: This is a measure of the ability of multiple actuators to work together for a common output. Because of the difference among the multiple actuator outputs, the force fighting is common and one should find a way to decouple the system and provide the compensation to drive the load together.

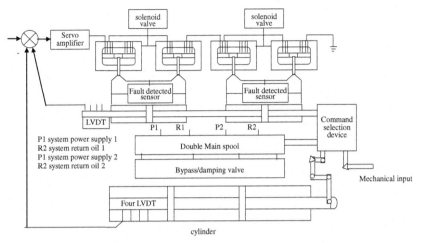

FIGURE 2.52   Electromagnetic synthesized redundant actuator.

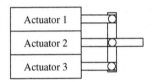

FIGURE 2.53   Force synthesized redundant actuator.

2. Force voting: Because of the difference among the multiple outputs of a redundant actuator, the force voting can provide the common output representing the mid value of all commands.

3. Active/standby: The active actuator is usually commanded by a single electronic channel, and mismatch is of no concern during operation. However, mismatches between the commands of the active and the standby channel are of concern and must be minimized to avoid large-surface transients upon switching from active to standby actuators.

4. Position summing: Position summing of secondary actuators results in a single output that is the average of the input commands.

2.4 PROOF TEST [26,27,28]

An aircraft hydraulic system must be qualified before the aircraft is approved for flight. The qualification is designed to verify that each individual component is meeting its specification. This includes proof and burst pressure tests, fatigue, vibration, acceleration, and functional tests. Because the proof tests are time-consuming, these may be complemented by accelerated life tests.

The entire hydraulic system is built up into a test rig. This rig includes a steel structure representing the aircraft into which the hydraulic piping and all components are mounted in their own position. The size and layout of the

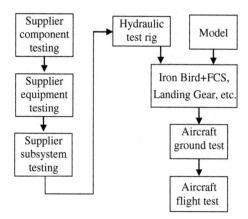

FIGURE 2.54 Hydraulic system testing process.

pipes are identical as in the aircraft. The standard pumps provide the flows and pressures similar to the aircraft hydraulic power supply system. The rig is equipped with load devices to simulate aerodynamic and other loads. A well-designed and operated hydraulic test rig is crucial to the process of the formal qualification and certification of the aircraft. A typical test rig is shown in Figure 2.54.

2.5 CONCLUSIONS

This chapter introduced different types of hydraulic systems in aircraft, such as the single, dual, and triplex hydraulic systems. The chapter provided analysis of the hydraulic systems of the Boeing and Airbus families of aircraft. The Boeing aircraft family uses triplex and quadruplex similar hydraulic systems whereas the Airbus family uses the dissimilar power supply systems to avoid common failure. For example, A380 chooses a 2H+2E architecture. This chapter also introduced the operational principle of typical hydraulic components, such as a hydraulic pump, reservoir, control valve, filter, pipe, and actuator. The chapter discussed design methods for hydraulic components. For the actuator, the static and dynamic design methods were introduced to satisfy the full performance requirement of HAs. Finally, the chapter provides the proof test of a hydraulic system.

REFERENCES

[1] I. Moir, A. Seabridge, Aircraft Systems: Mechanical, Electrical and Avionics Subsystems Integration, John Wiley & Sons, 2008.
[2] P.A. Stricker, Aircraft Hydraulic System Design, Eaton Aerospace Hydraulic System Division Report, 2010.

[3]   B. Hendrix, SAE ARP 4761 Process, Workshop AM Presentation, Safety Case Workshop.

[4]   M. Jelali, Andreas, Hydraulic servo-systems: modeling, identification and control, Springer, 2003.

[5]   Aerospace Recommended Practice, SAE International, 1996, ARP4761, 400 Commonwealth Drive, Warrendale, PA 15096-0001.

[6]   Aerospace Capabilities Boeing 777, Eaton Powering Business Worldwide, CS-21B B777_Capabilities.pdf.

[7]   A320 Hydraulic, Airbus Trainning: A320 Flight Crew Operating Manual, 1.29.00. http://www.smartcockpit.com/docs/A320-Hydraulic.pdf.

[8]   Aircraft Characteristics Airport and Maintenance Planning, Airbus A340-200/-300, Airbus S.A.S. Customer Services, Technical Data Support and Services, 31707 Blagnac Cedex, France. http://www.airbus.com/fileadmin/media_gallery/files/tech_data/AC/Airbus-AC-A340-200-300-20140101.pdf.

[9]   ARJ21 Hydraulic System RFP Response, Parker.

[10]   N.D. Manring, Hydraulic Control Systems, John Wiley & Sons, 2005.

[11]   Towards more electric aircraft, @Messier Bugatti, R3ASC04.

[12]   What Is a Flight Control System? WiseGEEK. http://www.wisegeek.com/what-is-a-flight-control-system.htm.

[13]   D. Crane, Dictionary of Aeronautical Terms, Aviation Supplies & Academics, 1997.

[14]   J. Gao, Z. Jiao, P. Zhang, Aircraft Fly-by-Wire System and Active Control Technique, Beijing University of Aeronautics and Astronautics Press, 2010 (in Chinese).

[15]   http://aerospace.eaton.com/news.asp?articledate=06/01/03&NewsCommand=ViewMonth.

[16]   http://www.tpub.com/content/aviation/14018/css/14018_178.htm.

[17]   M. Jelali, A. Kroll, Hydraulic Servo System, Springer, 1982, ISBN 1852336927.

[18]   W. Givens, P. Michael, in: G. Totten (Ed.), Fuels and Lubricants Handbook, ASTM International, 2003.

[19]   D. Placek, in: L. Rudnick (Ed.), Synthetics, Mineral Oils and Bio-Based Lubricants, CRC Press, 2006.

[20]   Bosch Automotive Handbook, fourth ed., Robert Bosch GmbH, 1996.

[21]   Discovery Channel-'Seconds from Disaster'. http://www.skydrol.com/pages/product.asp.

[22]   http://www.exxonmobil.com/USA-English/Aviation/PDS/GLXXENAVIEMExxon_HyJet_V.asp.

[23]   P. Dransfield, Hydraulic Control Systems—Design and Analysis of Their Dynamics, Springer-Verlag Berlin Heidelberg, New York, 1981.

[24]   FAR Part 25: Airworthiness Standards for Transport Category Airplanes, US Government Publishing Office. http://www.ecfr.gov/cgi-bin/text-idx?rgn=div5;node=14%3A1.0.1.3.11.

[25]   FAR Part 23: Airworthiness Standards for Normal, Utility, Acrobatic, and Commuter Category Airplanes, US Government Publishing Office. http://www.ecfr.gov/cgi-bin/text-idx?tpl=/ecfrbrowse/Title14/14cfr23_main_02.tpl.

[26]   FAR Part 21: Certification Procedures for Products and Parts, http://www.law.cornell.edu/cfr/text/14/part-21, AC 25.1309-1A System Design and Analysis Advisory Circular, 1998.

[27]   ARP4761: Guidelines and Methods for Conducting the Safety Assessment Process on Civil Airborne Systems and Equipment. http://standards.sae.org/arp4761/.

[28]   ARP 4754: Certification Considerations for Highly-integrated or Complex Aircraft Systems. http://standards.sae.org/arp4754/.

[29]   AIR5005: Aerospace—Commercial Aircraft Hydraulic Systems. http://standards.sae.org/air5005/.

[30] L.A. Johnson, DO-178: Software Considerations in Airborne Systems and Equipment Certification (incl. Errata Issued 3-26-99), Boeing Commercial Airplane Group, Seattle, Washington 98124-2207. http://www.dcs.gla.ac.uk/~johnson/teaching/safety/reports/schad.html.

[31] DO-254: Design Assurance Guidance for Airborne Electronic Hardware, Embedded Systems Guide. http://www.embedded-systems-portal.com/CTB/DO-254_Design_Assurance_Guidance_for_Airborne_Electronic_Hardware_10036.html.

[32] D.D. Ryder, Redundant Actuator Development Study, NASA CR-114730, Prepared under Contract NAS2-7653, Boeing Commercial Airplane Company, Seattle, Washington, 98124, 1973.

[33] B. Huang, Study on key techniques of aircraft intelligent variable pressure hydraulic system, PhD thesis of Beihang University, 2012.

[34] MIL-P-19692E, Military Specification: pumps, variable delivery, general specification, 1994.

[35] R.W. Pratt, Flight control systems: practical issue in design and implementation, The Institution of Electrical Enginees, 2000.

Chapter 3

Comprehensive Reliability Design of Aircraft Hydraulic System

3.1 QUALITY AND RELIABILITY

Product quality and reliability are commonly considered to be synonymous. However, there is a difference between the two words. The quality of the component or system generally refers to whether a component or a system meet performance specification requirements; it does not reflect how long the product continues to meet specifications. On the other hand, the reliability is the ability of a system or component to perform its required functions under specified conditions for a period of time. Hence, the reliability can be considered as the change in quality over time, and their difference is the useful time. Quality is a characteristic of a product at the start of its life, whereas reliability is its characteristic at any particular moment in time.

When the product design requirement changes from pure performance to high efficiency, the scope of quality also changes. The modern approach to product development requires that the product quality should meet the integrated characteristics, including performance, reliability, maintainability, safety, testability, supportability, and economics, Figure 3.1. Herein, performance, reliability, maintainability, testability, supportability, and safety are the primary quality characteristics and economic is the secondary quality characteristic that means the product should have a reasonable lifecycle cost.

Commercial Aircraft Hydraulic Systems. http://dx.doi.org/10.1016/B978-0-12-419972-9.00003-6

FIGURE 3.1   Integrated quality characteristics.

The product quality depends on the comprehensive evaluation based on the above characteristics, especially on the reliability and performance. Here, we apply a term—*comprehensive reliability*—to include all of the above characteristics.

In the beginning of product design, the designer focused on the performance and ignored the influence of reliability. In 1979, a DC-10 commercial aircraft crashed and all passengers were killed. Later, it was found that the reason was that the repair process was not following the guidelines, which caused excessive stress that cracked the engine mounts. Therefore, it is necessary to consider the reliability, maintainability, testability, supportability, and performance during the product design phase.

The aircraft hydraulic system is a most important system that provides the high-pressure fluid for aircraft takeoff, maneuverability, undercarriage control, and antiskid brakes. Because the aircraft hydraulic system is complex, it is necessary to design it with high reliability, maintainability, and safety.

3.2 COMPREHENSIVE RELIABILITY

3.2.1 Development of Aircraft-related Reliability

From 1930 to 1940, people thought a product's useful life was deterministic; therefore, preventive maintenance was widely used in equipment maintenance. At that time, they believed that every product would wear and tear, fatigue, or age when it was put into use. The fatigued material could be replaced and structural repair performed when the design life was achieved. In the 1950s, US scientists discovered that things were not so simple. Using scheduled maintenance could not significantly reduce the failures in many cases despite the increased number of repair times. Scientists calculated the use of weapons in World War II and discovered that every $1 spent on equipment needed $2 to be spent in maintenance. This means that there are many weapons that could not be used and needed to be repaired during the war. The repair technician was very busy checking and disassembling the fault item, and such big repair effort did not decrease the frequency of faults. Therefore, became apparent that basic research in the area of reliability is necessary to support design process and product life-cycle. In 1952, the US military organized the Advisory Group on Reliability of Electronic

Equipment (AGREE) to study the reliability principle and redefine reliability [1,2]. After several years of research AGREE came up with the following findings: (1) electronic components had a failure rate which was related to its material, manufacturing process, and operational environment; and (2) the component failure distribution can be obtained with fault statistics. In 1957, AGREE first gave the reliability definition; namely, reliability is the probability that the product will operate without failure under fixed operational time, and the reliability discipline was born [3]. To specify the reliability procedure, AGREE started to formulate the standards. In 1962, the first reliability standard, MIL-STD-721A, was published. From 1960, development of reliability field was very fast and some common methods appeared, such as reliability block diagram (RBD), failure mode effect analysis (FMEA), fault tree analysis (FTA), and the Markov state transfer method. In the 1970s and 1980s, work on development of software for reliability analysis resulted in number of different reliability models and analysis methods. In 1990, the computer network appeared and network reliability analysis became the hot topic in the reliability field.

The development of aeronautics and astronautics greatly pushed the study of reliability forward. In the early period of aircraft development aircraft focus was designed behind the center of gravity, which was called the *static stability design method*. In this concept, when an aircraft encounters wind disturbance, it can automatically return to steady state. However, the static stability design degraded maneuverability. To overcome this problem, the static instability design method was created, in which the aircraft focus was designed to be in front of the center of gravity. In this case, the aircraft was not stable at low speeds. The problem of maintaining aircraft stability was a very important issue in aircraft design. The active control technique (ACT) was used to improve system stability through the control method. Because the focus of aircraft would move backward with increasing aircraft speed, the aircraft would be stable at high speeds. ACT can guarantee aircraft stability through increasing stability augmentation and control augmentation at low speeds. However, the ACT consists of sensors, controllers, and actuators, which are not reliable; therefore, it is necessary to design redundancy systems to achieve high reliability. Other safe critical systems have the same problem. Reliability is currently designed and analyzed throughout the product lifecycle by companies such as Ford Motor Company and Boeing Company as well as the National Aeronautics and Space Administration [5]. The conventional reliability theory is based on probability.

3.2.2 Definition of Comprehensive Reliability

Comprehensive reliability consists of five characteristics: reliability, maintainability, safety, testability, and supportability.

- Reliability: This is the ability of a system or a component to perform its required functions under stated conditions for a specified period of time. Reliability depends on operational and environmental conditions. Reliability is a function of time.
- Maintainability: This is the probability that a system or component will be retained in or restored to a specified condition within a given period of time, when the maintenance is performed in accordance with prescribed procedures and resources. If the reliability and maintainability are considered at the same time, availability is used to express its characteristics. Availability is the probability that a system or component is available whether scheduled or not.
- Safety: Safety is the condition of being protected against failure, damage, accidents, or any other event that could be considered detrimental to the system operation. Safety can also be defined as the control of recognized hazards to achieve an acceptable level of risk. For the aircraft, this can take the form of being protected from the event that causes an aircraft to crash or to result in financial losses. Aircraft safety hazards include foreign object debris, erroneous information, lightning, ice and snow, engine fire, structure damage, stalling, fire, bird strike, etc.
- Testability: Testability is a characteristic that indicates whether a system or component can accurately determine its status (operation, failure, or performance degradation) and isolate its internal failure in a timely manner.
- Supportability: This is a characteristic that a system or component can provide logistics support over its entire lifecycle. The main factors of supportability include the current health status of the system, required maintenance events, support resources needed, and rapid repair management.

3.3 COMPREHENSIVE RELIABILITY THEORY

3.3.1 Reliability Theory of a Nonrepairable System

Because reliability focuses on random events, its theory is based on probability and statistics. According to the military standard, MIL-STD-721 [3], the definition of failure is "the event or inoperable state, in which any item or part of an item does not, or would not, perform as previously specified."

The opposite of failure is *reliability*, which is "the ability of a system or component to perform its required functions under stated conditions for a specified period of time." For nonredundant items, the reliability is the duration or probability of failure-free performance under stated conditions. For redundant items, the reliability is the mission reliability. There are four parameters (discussed in the following sections) to describe the product reliability under a nonrepairable system.

3.3.1.1 Reliability Function

According to the definition, reliability is the probability that a product will operate properly for a specified period of time under the designed operating conditions [6]. Reliability is a function of time t and can be described as

$$R(t) = \begin{cases} P(T > t) & (t \geq 0) \\ 1 & (t < 0) \end{cases} \tag{3.1}$$

where T is the life of the item, which is a random variable, and t is time. The operating condition consists of environmental conditions (temperature, humidity, vibration, etc.) and operational conditions (pressure, velocity, load, voltage, etc.). When the failure distribution function of T is known as $f(t)$, the reliability can be described as

$$R(t) = \begin{cases} \displaystyle\int_{t}^{\infty} f(t)\mathrm{d}t & (t \geq 0) \\ 1 & (t < 0) \end{cases} \tag{3.2}$$

The reliability function is represented in Figure 3.2, in which the reliability of the item decreases from 1 to 0 as the time increases.

In some applications, we focus on mission reliability, which is the ability of a product to perform its required functions for the duration of a specified "mission profile."

3.3.1.2 Cumulative Probability of Failure

Cumulative probability of failure is the probability that the product will fail under fixed operational time, which is denoted with $F(t)$.

$$F(t) = \begin{cases} P(0 < T \leq t) & (t \geq 0) \\ 0 & (t < 0) \end{cases} \tag{3.3}$$

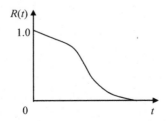

FIGURE 3.2   Reliability versus time.

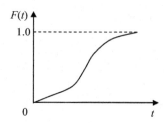

FIGURE 3.3   Failure probability versus time.

When the failure distribution function of product life T is known as $f(t)$, $F(t)$ can be described as

$$F(t) = \begin{cases} \displaystyle\int_0^t f(t)\mathrm{d}t & (t \geq 0) \\ \quad 0 & (t < 0) \end{cases} \qquad (3.4)$$

If $F(t)$ is continuous then

$$f(t) = \frac{\mathrm{d}F(t)}{\mathrm{d}t} \quad (t \geq 0) \qquad (3.5)$$

From Eqns (3.2) and (3.4), we obtain

$$R(t) + F(t) = 1 \qquad (3.6)$$

The cumulative probability of the failure curve can be described as shown in Figure 3.3, in which the cumulative probability of failure increases with operational time from 0 to 1.

3.3.1.3 Failure Rate λ(t)

Failure rate is the limit of the probability that a failure occurs per unit time interval Δt given that no failure has occurred before time t. The failure rate is the conditional probability, which can be expressed as

$$\lambda(t) = \lim_{\Delta t \to 0} \frac{P(t < T \leq t + \Delta t | T > t)}{\Delta t} = \frac{\lim_{\Delta t \to 0}\{F(t + \Delta t) - F(t)\}/\Delta t}{R(t)} = \frac{f(t)}{R(t)} \qquad (3.7)$$

The traditional bathtub curve can describe the variance of failure rate shown in Figure 3.4.

Figure 3.4 shows the bathtub curve of a nonrepairable product, in which the first part shows a decreasing failure rate, known as *early failure*; the second part is a constant failure rate, known as *random failure*; and the third part is an increasing failure rate, known as *wear-out failure*. In general, a product's failure rate is high in the beginning operation because of early failure of components. Its failure rate will decrease very fast when a defective component of the

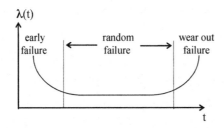

FIGURE 3.4   Bathtub curve of failure rate [6].

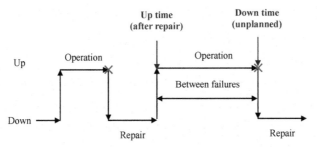

FIGURE 3.5   The operational process of the repairable product.

product is identified and discarded. After the early failures are eliminated, the product enters a steady operational condition with a low and constant failure rate. During this period, the failures are caused by random factors. In the later period of life of the product, the failure rate increases with product's maturing age caused by progressive wear and tear. Most of the product lifecycle behaves according to the bathtub curve.

3.3.1.4 Mean Time to Failure and Mean Time between Failures

According to the maintenance characteristics, a product can be characterized into two types: a nonrepairable product or a repairable product. Herein, the mean time to failure (MTTF) is used to describe the mean life of a non-repairable product, and the mean time between failure (MTBF) is used to express the mean life of a repairable product.

MTTF is the arithmetic mean time to failure of a product; therefore, it can be described as

$$MTTF = \int_0^\infty tf(t)dt = \int_0^\infty R(t)dt \tag{3.8}$$

When a product can be repaired, the operational process of the product can be represented as shown in Figure 3.5.

MTBF is the arithmetic mean between failures of a product during operation, and it can be described as

$$MTBF = \frac{\sum_{i=1}^n t_i}{n} \tag{3.9}$$

where t_i is the operational time between failures of a repairable product and n is the number of repairs. When the failure rate is constant, *MTBF* can be expressed as

$$MTBF = \frac{1}{\lambda} \qquad (3.10)$$

MTBF is a very important reliability measure in safety-critical systems. In addition to characterizing the useful time of a product, MTBF also governs the frequency of required system maintenance and inspections.

3.3.2 Reliability Theory of a Repairable System

In real application, a repairable product requires maintenance and changes in faulty items to maintain the operational state. Therefore, the product lifecycle alternately goes through normal and faulty states, as shown in Figure 3.5. Figure 3.5 shows that the lifecycle of a repairable product includes an operational period and a repair period, in which the product reliability index measures its statistical regulation in the operational period and the maintainability index expresses its regulation in the repair period.

3.3.2.1 Maintainability M(t)

Maintainability is the measure of the ability of an item to be retained in or restored to a specified condition when maintenance is performed by personnel having specified skill levels using prescribed procedures and resources at each prescribed level of maintenance and repair.

Maintainability depends on the product design and the technical level of repair personnel, the repair process, and the repair facilities. Maintainability is a probability measure that a product will be maintained or restored to the specified function for a specified period of time at any specific level of repair conditions.

Suppose that T_m is the repair time with repair density function $m(t)$. The product maintainability can then be expressed as

$$M(t) = \begin{cases} P(0 < T_m \leq t) = \int\limits_0^t m(t)\mathrm{d}t & (t \geq 0) \\ \\ 0 & (t < 0) \end{cases} \qquad (3.11)$$

If $M(t)$ is differentiated, then

$$m(t) = \frac{\mathrm{d}M(t)}{\mathrm{d}t} \quad (t \geq 0) \qquad (3.12)$$

Therefore, the repair density function $m(t)$ is the probability that the faulty product is repaired to normal condition in Δt.

3.3.2.2 Repair Rate μ(t)

The repair rate is the conditional probability that the product is repaired to specified function in $(t, t+\Delta t)$ when the product fails at time t. The repair rate can be described as

$$\mu(t) = \lim_{\Delta t \to 0} \frac{P(t < T_m \leq t + \Delta t | T_m > t)}{\Delta t} = \frac{m(t)}{1 - M(t)} \qquad (3.13)$$

3.3.2.3 Mean Time to Repair

MTTR is a basic measure of the maintainability of a repairable product. It represents the mathematical expectation of a failed product's repair time:

$$MTTR = E(T_m) = \int_0^\infty tm(t)\mathrm{d}t \qquad (3.14)$$

3.3.2.4 Availability A(t)

Availability is a measure of the degree to which a product is in an operable state at the start of a mission when the mission is called for at a random time. Availability is the integrated parameter considering the operational and repair conditions. If the reliability and maintainability of a product follow an exponential distribution (i.e., $R(t) = e^{-\lambda t}$, $M(t) = 1 - e^{-\mu t}$), then transient availability can be described as follows:

$$A(t) = \frac{\mu}{\lambda + \mu} + \frac{\lambda}{\lambda + \mu} e^{-(\lambda + \mu)t} \qquad (3.15)$$

When $\mu = 0$, then

$$A(t) = R(t) = e^{-\lambda t} \qquad (3.16)$$

When $t \to \infty$, the limit of availability exists, and the steady availability can be expressed as

$$A = \frac{\mu}{\lambda + \mu} = \frac{MTBF}{MTBF + MTTR} \qquad (3.17)$$

It is obvious that steady availability not only can describe the product reliability parameter with failure rate λ (or MTBF) but also the maintainability parameter with repair rate μ (or MTTR).

3.3.3 Conventional Probability Density Function

In reliability analysis, the probability density function of product is a very important measure to describe the failure distribution. Different products have different failure mechanisms and thus different failure distributions. Through product life testing, it is easy to determine the product probability density function. The conventional statistical distribution includes the exponential distribution, Weibull distribution, and normal distribution.

3.3.3.1 Exponential Distribution

Exponential distribution is widely used in electrical products, and its probability density function can be described as

$$f(t) = \lambda e^{-\lambda t} \tag{3.18}$$

where $t \geq 0$. The reliability of the exponential distribution is

$$R(t) = \int_{t}^{\infty} \lambda e^{-\lambda t} \mathrm{d}t = e^{-\lambda t} \tag{3.19}$$

The reliability under exponential distribution decreases very fast with increases in the operational time, Figure 3.6(a). Hence, it is necessary to find a way to improve the reliability of electrical products.

The failure rate under exponential distribution is constant, as follows:

$$\lambda(t) = \frac{f(t)}{R(t)} = \frac{\lambda e^{-\lambda t}}{e^{-\lambda t}} = \lambda = \text{constant} \tag{3.20}$$

The mean time under exponential distribution is the reciprocal of the failure rate, as follows:

$$\theta(MTTF \text{ or } MTBF) = \int_{0}^{\infty} tf(t)\mathrm{d}t = \frac{1}{\lambda} \tag{3.21}$$

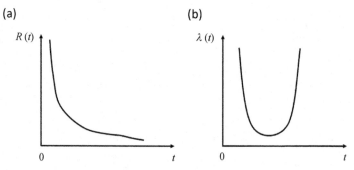

FIGURE 3.6 Reliability and failure rate versus time for exponential distribution. (a) Reliability of exponential distribution, (b) failure rate of exponential distribution.

There is a very important characteristic in exponential distribution—namely, memorylessness. It can be described with the following conditional probability:

$$P\{(T > t_0 + t)|(T > t_0)\} = P(T > t) \qquad (3.22)$$

Memorylessness (aka *evolution without after-effects*), a measure of the number of arrivals occurring at any bounded interval of time after time t, is independent of the number of arrivals occurring before time t. The proof of this relationship is follows.

Proof: According to the conditional probability, Eqn (3.22) can be expressed as

$$P\{(T > t_0 + t)|(T > t_0)\} = \frac{P(T > t_0 + t) \cap P(T > t_0)}{P(T > t_0)}$$

$$= \frac{P(T > t_0 + t)}{P(T > t_0)} = \frac{e^{-\lambda(t_0+t)}}{e^{-\lambda t_0}} = e^{-\lambda t} \qquad (3.23)$$

$$= P(T > t)$$

Therefore, Eqn (3.22) is proved.

3.3.3.2 Normal Distribution

Normal distribution (or Gaussian distribution) is a very commonly occurring continuous probability distribution. It is a function that tells the probability that any real observation will fall between any two real limits. Normal distributions are extremely important in statistics and are often used in the natural and social sciences for real-valued random variables in unknown distributions. The normal distribution is very useful because of the central limit theorem, which states that, under normal conditions, the mean of many random variables independently drawn from the same distribution is distributed approximately normally, irrespective of the form of the original distribution. Physical quantities that are expected to be the sum of many independent processes (e.g., measurement errors) often have a distribution very close to normal. Moreover, many results and methods (e.g., propagation of uncertainty and least-squares parameter fitting) can be derived analytically in explicit form when the relevant variables are normally distributed.

A normal distribution is defined as

$$f(t) = \frac{1}{\sqrt{2\pi}\sigma} e^{-\frac{1}{2}\left(\frac{t-\mu}{\sigma}\right)^2} \qquad (3.24)$$

where μ is the mean or expectation of the distribution, σ is its standard deviation, and σ^2 is the variance. A random variable with a Gaussian distribution is said to be normally distributed and is called a *normal deviation*. Here,

$$MTTF(\text{or } MTBF) = E(T) = \mu \qquad (3.25)$$

If $\mu = 0$ and $\sigma = 1$, then the distribution is called the *standard normal distribution* or the *unit normal distribution*, and a random variable with that distribution is a *standard normal deviate*. To simplify calculations, we commonly use the unit normal distribution. Let $Z = (t - \mu)/\sigma$; then, the normal distribution can be transferred into unit normal distribution as follows:

$$F(Z) = \phi(Z) = \frac{1}{\sqrt{2\pi}} \int_{-\infty}^{Z} e^{-\frac{Z^2}{2}} dZ \qquad (3.26)$$

The cumulative failure probability of unit normal distribution can be obtained according to Table A.1 (Appendix A).

3.3.3.3 Weibull Distribution

Weibull distribution is widely used to describe the failure of a mechanical product. There are three parameters in Weibull distribution: shape parameter m, scale parameter t_0, and location parameter γ. The probability density function of Weibull distribution can be written as follows:

$$f(t) = \frac{m}{t_0}(t - \gamma)^{m-1} e^{-\frac{(t-\gamma)^m}{t_0}} \quad (\gamma \le t, 0 < m, t_0) \qquad (3.27)$$

The cumulative distribution function of Weibull distribution is

$$F(t) = 1 - e^{-\frac{(t-\gamma)^m}{t_0}} \qquad (3.28)$$

The reliability function of Weibull distribution is

$$R(t) = e^{-\frac{(t-\gamma)^m}{t_0}} \qquad (3.29)$$

The failure rate of Weibull distribution is

$$\lambda(t) = \frac{f(t)}{R(t)} = \frac{m}{t_0}(t - \gamma)^{m-1} \qquad (3.30)$$

The *MTTF* can be written as follows:

$$MTTF(MTBF) = E(T) = t_0^{\frac{1}{m}}\Gamma\left(1 + \frac{1}{m}\right) = \eta\Gamma\left(1 + \frac{1}{m}\right) \qquad (3.31)$$

where $\eta = t_0^{1/m}$ is the characteristic life described by Weibull distribution and Γ is the gamma function.

3.4 RELIABILITY DESIGN OF A HYDRAULIC SYSTEM

Aircraft hydraulic system design requires that aircraft must maintain control under all normal and anticipated failure conditions. Many system

TABLE 3.1 Aircraft Hydraulic System Safety Standards

Failure Criticality	Failure Characteristics	Probability of Occurrence	Design Standard
Single hydraulic system fails	Normal, nuisance, and/or possibly requiring emergency procedures	Reasonably probable	NA
Two (out of three) hydraulic systems fail	Reduction in safety margin, increased crew workload, may result in some injuries	Remote	$P \leq 10^{-5}$
All hydraulic sources fail except RAT or auxiliary power unit (APU)	Extreme reduction in safety margin, extended crew workload, major damage to aircraft, and possible injury and deaths	Extremely remote	$P \leq 10^{-7}$
Catastrophic	Loss of aircraft with multiple deaths	Extremely improbable	$P \leq 10^{-9}$

NA, not applicable.

architectures and design approaches are developed to meet this high-level requirement. Hydraulic system architecture can arrange and interconnect hydraulic power sources and consumers for controllability of the aircraft. Hence, the components should be designed with reliability design methods and systems should be designed with redundancy technology in the case of potential failures. Aircraft hydraulic system safety standards are shown in Table 3.1.

3.4.1 Reliability Design of Electrical Components

Since electrical components (linear variable differential transformer (LVDT) or other electronic component) in a hydraulic system are sensitive to the environment, the reliability design for this kind of component focuses on the environmental adaptability design. Analysis of the failure data of electrical components, indicates that several factors can cause electrical component failure include electromigration, drift of component parameters, transient electrical stresses, excessive heat, and electromagnetic inference (EMI). To maintain high reliability of electrical components, an electrical component should be designed within the required characteristics, such as temperature, vibration, humidity, EMI, and parameter error range. Before the electrical component is used in real applications, it requires a reliability test under temperature cycle, random vibration, and humidity.

An example of the reliability design for an electrical component is presented in the following subsections.

3.4.1.1 Tolerance Design

Suppose there are several design parameters $p_1, p_2, ..., p_n$ in an electrical product's design, and that its performance parameter V_i can be described as

$$V_i = f(p_1, p_2, ..., p_n) \quad (i = 1, 2, ..., m) \tag{3.32}$$

The Taylor expansion of Eqn (3.32) can be described as

$$\Delta V_i = \sum_{j=1}^{n} \frac{\partial V_i}{\partial p_j}\bigg|_0 \Delta p_j \tag{3.33}$$

where ΔV_i is the variance of V_i and Δp_i is the variance of the ith design parameter. The subscript "0" indicates the nominal value.

The sensitivity of the product performance characteristics to design parameter deviations is defined as

$$S_{ij} = \frac{\Delta V_i / V_{i_0}}{\Delta p_j / p_{j_0}} \tag{3.34}$$

We can then obtain the performance characteristics of the corresponding value of deviation sensitivity as

$$\Delta V_i / V_{i_0} = \sum_{j=1}^{n} S_{ij} \Delta p_j / p_{j_0} \tag{3.35}$$

Using Eqn (3.35), we can design an electrical product within the parameters' tolerance.

3.4.1.2 Thermal Design

In general, every component operates at a certain temperature range. When temperature exceeds the design limit, the physical performance of material will change and its failure rate will increase, as shown in Figure 3.7 [6].

In general, electronic devices can easily fail at high and low temperatures; therefore, it is necessary to make the product operate within a design limit so as to maintain high reliability. There are three ways to maintain temperature of the component within desired temperature range: conduction, convection, and radiation. Adapting the appropriate cooling system, will allow electronic component to dissipate excess heat. Appropriate installation and layout of the components can decrease the heat generation and propagation.

In real applications, different components have different reliability design methods. For example, the reliability design of an outdoor electrical

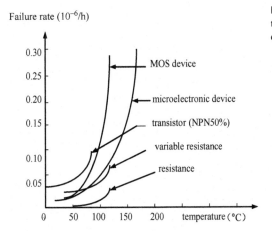

Failure rate (10⁻⁶/h)

component should cover the waterproof design, cable design, antivibration design, etc., as shown in Figure 3.8 [6].

3.4.2 Reliability Design of a Mechanical Product

The common reliability design of a mechanical product relies on its mechanical properties and stress distribution, where both stress and strength use their respective probability distributions. The major difference between an electrical or electronic component and a mechanical component is that the failure of the mechanical component is sensitive to operational conditions. A mechanical component is considered to be reliable when the strength of the component exceeds stresses caused by the load acting on it [17,18]. Suppose the random variable δ is the material strength and the random variable s is the stress. Their probability density functions are then described as $g(\delta)$ and $f(s)$, respectively. Define the reliability of mechanical component as

$$R = P(\delta > s) = P(\delta - s > 0) \tag{3.36}$$

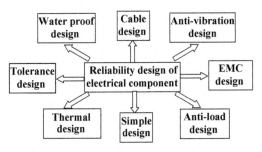

FIGURE 3.8 Reliability design of electrical components.

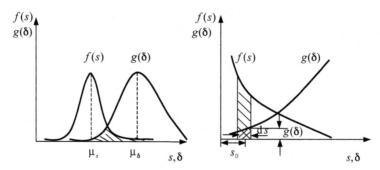

FIGURE 3.9   Stress-strength interference curve [6].

Suppose that δ and s are independent variables with normal distribution. The cumulative failure probability can then be calculated with overlapping area as shown in Figure 3.9.

For any stress s_0, enlarge the interference area, and then the probability that the stress falls in the ds is equal to the area of unit ds:

$$P\left(s_0 - \frac{ds}{2} \leq s \leq s_0 + \frac{ds}{2}\right) = f(s_0)ds \tag{3.37}$$

The probability that the strength δ is greater than s_0 can be described as

$$P(\delta > s_0) = \int_{s_0}^{\infty} g(\delta)d\delta \tag{3.38}$$

Then, the probability that the strength is greater than the stress is the reliability of the component:

$$R = \int_{-\infty}^{\infty} f(s)\left[\int_{s}^{\infty} g(\delta)d\delta\right]ds \tag{3.39}$$

Conversely, the reliability of a component can also be derived according to the condition that the stress is less than the strength, as follows:

$$R = \int_{-\infty}^{\infty} g(\delta)\left[\int_{-\infty}^{\delta} f(s)ds\right]d\delta \tag{3.40}$$

Equation (3.39) is similar to Eqn (3.40).

If the stress s has a normal distribution,

$$f(s) = \frac{1}{\sigma_s\sqrt{2\pi}}\exp\left[-\frac{1}{2}\left(\frac{s - \mu_s}{\sigma_s}\right)^2\right] \tag{3.41}$$

and if strength δ also has a normal distribution,

$$g(\delta) = \frac{1}{\sigma_\delta \sqrt{2\pi}} \exp\left[-\frac{1}{2}\left(\frac{\delta - \mu_\delta}{\sigma_\delta}\right)^2\right]$$

(3.42)

$$(-\infty < \delta < \infty)$$

where μ_s and σ_s are the mean value and standard deviation, respectively, of stress s, and μ_δ and σ_δ are the mean value and standard deviation, respectively, of strength δ. Then, define a new random variable y as

$$y = \delta - s$$

(3.43)

The reliability of the component can then be described as [16]

$$R = P(y > 0) = \int_0^\infty \frac{1}{\sigma_y \sqrt{2\pi}} \exp\left[-\frac{1}{2}\left(\frac{y - \mu_y}{\sigma_y}\right)^2\right] dy$$

(3.44)

where $\mu_y = \mu_\delta - \mu_s$ is the mean value of random variable y and $\sigma_y = \sqrt{\sigma_\delta^2 + \sigma_s^2}$ is the standard deviation of random variable y.

Example: Consider an axial piston in a hydraulic pump consisting of a cylinder barrel, pistons, piston shoes, a cylinder barrel-valve plate friction pair, a piston-cylinder barrel friction pair, and a pressure-regulating device, as shown in Figure 2.12. To evaluate the reliability of the hydraulic pump, the stress-strength interface model is used to calculate the reliability of the hydraulic pump.

Solution:

1. Fatigue reliability of the cylinder barrel

The cylinder barrel is the main component of the hydraulic pump, in which the pistons reciprocate within the plunger cavity.

● Material parameters

The material of the cylinder barrel is steel and the tensile strength is μ_B. Then, the mean value and standard deviation of the tensile strength are

$$\mu_{\delta_B} = \frac{\mu_B}{1 - 1.282C}$$

$$\sigma_{\delta_B} = \mu_{\delta_B} C$$

(3.45)

where $C = 0.075$ [6] is the material variation coefficient. Here, $\mu_{\delta_B} = 968$ MPa, $\sigma_{\delta_B} = 72.7$ MPa. According to the mechanical manual, the bending fatigue limit of the material is $\delta_{-1} = 270$ MPa, and the mean value and standard deviation are written as

$$\mu_{\delta_{-1}} = \frac{\delta_{-1}}{1 - 1.282C} = 309.7 \text{ MPa}$$

$$\sigma_{\delta_{-1}} = \mu_{\delta_{-1}} C_{\delta_{-1}} = 10.97 \text{ MPa}$$

(3.46)

(a)   (b)

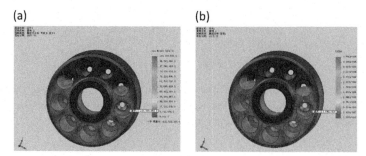

FIGURE 3.10   Stress distribution of a cylinder barrel. (a) Stress distribution, (b) strain distribution.

- Stress distribution of the cylinder barrel

Considering the real operating conditions, the main load on the cylinder barrel is due to the high-pressure fluid. Figure 3.10 shows the stress distribution and strain distribution of the cylinder barrel determined by finite element analysis (FEA) method.

In Figure 3.10, it is obvious that the maximum stress point is at the transition area from high pressure to low pressure, approximately 105.3 MPa. The tensile strength of this steel is $\mu_B = 620$ MPa; therefore, the strength of the material is sufficient to resist the maximum stress. The maximum strain at the plunger cavity wall between high pressure and low pressure of the cylinder barrel is 3.496e-4; therefore, the strain of the cylinder is in the acceptable range.

According to the simplified model of the thick wall cylinder, Figure 3.11, and elasticity theory, the maximum stress of the cylinder bore wall can be obtained considering the interference between holes:

$$S_{max} = p_S \sqrt{\frac{4(n^2+1)^2}{(n^2-1)^2} - 2\left(1+\frac{1}{m}\right)\left(\frac{n^2+3}{n^2-1}\right)} = 3.136 p_S = 109.74 \text{ MPa}$$

(3.47)

where the rated pressure $p_S = 35$ MPa and $\delta_{max} = \frac{0.1 S_{max}}{3} = 3.659$ MPa.

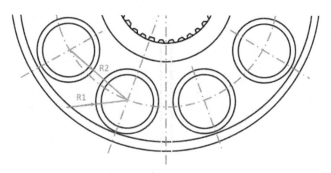

FIGURE 3.11   Simplified model of a thick wall cylinder [6].

The mean stress value and standard deviation can be obtained as follows

$$\mu_s = \frac{\sqrt{2}}{2} S_{\max} = 77.6 \text{ MPa}$$

(3.48)

$$\sigma_s = 0.5 \delta_{\max} = 1.83 \text{ MPa}$$

- Fatigue strength distribution of the cylinder barrel

Let $b = \frac{\mu_{\delta_B}^2}{\mu_{\delta_{-1}}} = 3025.6$ MPa and $tg\theta = \frac{1-r}{1+r} = 1$; the mean strength and standard deviation of the cylinder barrel material can then be written as

$$\mu_{\delta_m} = \frac{-tg\theta b + \sqrt{(tg\theta b)^2 + 4\mu_{\delta_B}^2}}{2} = 283.2 \text{ MPa}$$

(3.49)

$$\sigma_{\delta_m} = \frac{\mu \delta_{\lim} - \delta_{\lim}}{3} = 50.7 \text{ MPa}$$

Considering the stress concentration, the strength mean value and standard deviation of the material can be written as

$$\mu_{\delta_K} = \frac{\mu_{\delta_{\lim}}}{K_\delta} = 160.2 \text{ MPa}$$

(3.50)

$$\sigma_{\delta_K} = \frac{\sigma_{\delta_{\lim}}}{K_\delta} = 20.3 \text{ MPa}$$

where $K_\delta = 2.5$ is the coefficient of stress concentration.

- Fatigue reliability of the cylinder barrel

Because $Z = \frac{\mu_{\delta_K} - \mu_s}{\sqrt{\sigma_{\delta_K}^2 + \sigma_s^2}} = 4.05$, the reliability of the cylinder barrel can be obtained from Table 3.1:

$$R_1 = 0.999974$$

(3.51)

Likewise, the other mechanical components' reliability values can be obtained if the stress distribution and material strength distribution are known.

2. Piston fatigue reliability

Suppose the piston material is 38CrMoAl. The material has the following mechanical properties: mean value of tensile strength $\mu_{\delta_B} = 1106$ MPa, standard deviation of tensile strength $\sigma_B = 83$ MPa, mean value of fatigue strength $\mu_{\delta_{-1}} = 440$ MPa, standard deviation of fatigue strength $\sigma_{\delta_{-1}} = 33$ MPa. The geometric parameters are piston diameter $d' = 20$ mm, and maximum angle of the swash plate is $\gamma = 17°$.

- Stress distribution of the piston

Through the FEA modeling the stress distribution and piston strain distribution are shown in Figure 3.12 under real operating conditions.

(a)   (b)

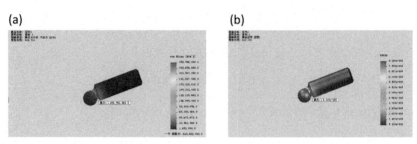

FIGURE 3.12 Piston stress and strain distribution. (a) Stress distribution of piston, (b) strain distribution of piston.

Both maximum stress and maximum strain are at the neck section of the piston. The maximum stress in the piston is 254.8 MPa. If the piston material is 38CrMoAl, then the tensile strength is $\mu_{\delta_B} = 1106$ MPa, which is sufficient to withstand stresses in the piston. The maximum strain is 0.234e-004, which means that the piston strain is very small and within the acceptable range.

The weak part of the piston is at the intersection between the piston head and piston AA', as shown in Figure 3.13.

In general, take the pressing force coefficient of the piston shoes friction pair as 1.06; then, the axial force of the piston is $F_p = 1.06pA$ and the radial force of the piston is $F_r = 0.205F_p$. The mean value and standard deviation due to the axial force F_p are

$$s_1 = \frac{1.06pA}{A'} = 108.14 \text{ MPa}$$

$$\sigma_{s_1} = \frac{0.1s_1}{3} = 3.6 \text{ MPa}$$

The mean value and standard deviation due to the radial force F_r are

$$s_{max} = \frac{F_p \dfrac{D}{2}}{0.1d'^3} = 148.57 \text{ MPa}$$

$$\sigma_{s_{max}} = \frac{0.1s_{max}}{3} = 4.95 \text{ MPa}$$

Then, the mean value and standard deviation of the maximum stress at the piston neck section are

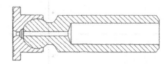

FIGURE 3.13 The intersection of a piston [6].

$$\mu_{s_{max}} = s_1 + s_{max} = 256.7 \ \text{MPa}$$

$$\sigma_{s_{max}} = \sqrt{\sigma_{s_1}^2 + \sigma_{s_{max}}^2} = 6.12 \ \text{MPa}$$

According to the general principle, the mean value and standard deviation of the operational stress are

$$\mu_s = \frac{\sqrt{2}}{2} \mu_{s_{max}} = 181.52 \ \text{MPa}$$

$$\sigma_s = \frac{1}{2} \sigma_{s_{max}} = 3.06 \ \text{MPa} \tag{3.52}$$

- Piston strength distribution

Let $b = \dfrac{\mu_{\delta_B}^2}{\mu_{\delta_{-1}}} = 2780.1 \ \text{MPa}$. The average strength of the piston can then be calculated as

$$\delta_m = \frac{-\tan \theta b + \sqrt{(\tan \theta b)^2 + 4\left(\mu_{\delta_B} - 3\sigma_{\delta_B}\right)^2}}{2} = 242.95 \ \text{MPa}$$

Then, the mean value of the piston strength can be written as

$$\mu_{\delta_m} = \frac{-\tan \theta b + \sqrt{(\tan \theta b)^2 + 4\mu_{\delta_B}^2}}{2} = 386.3 \ \text{MPa}$$

The mean value and standard deviation of the piston fatigue limit are

$$\mu_{\delta_{\text{lim}}} = \sqrt{\mu_{\delta_m}^2 + \left(\mu_{\delta_m} tg\theta\right)^2} = 546.25 \ \text{MPa}$$

$$\sigma_{\delta_{\text{lim}}} = \frac{\mu_{\delta_{\text{lim}}} - \delta_{\text{lim}}}{3} = 50.7 \ \text{MPa}$$

Let $K_\delta = 1.8$. Then, the mean value and standard deviation of the piston fatigue strength are

$$\mu_{\delta_{\text{lim}K}} = \frac{\mu_{\delta_{\text{lim}}}}{K_\delta} = 300.5 \ \text{MPa}$$

$$\sigma_{\delta_{\text{lim}K}} = \frac{\sigma_{\delta_{\text{lim}}}}{K_\delta} = 29.3 \ \text{MPa} \tag{3.53}$$

- Piston fatigue reliability

Because $Z = \dfrac{\mu_{\delta_{\text{lim}K}} - \mu_s}{\sqrt{\sigma_{\text{lim}K}^2 + \sigma_s^2}} = 3.995$, the fatigue reliability of the piston can be obtained from Table A.1 (Appendix A).

$$R_2 = 0.999967 \tag{3.54}$$

3. Wear reliability of the piston shoe

Let's assume that the material of a piston shoe is HMn58, with a mean value and standard deviation of its strength being $\mu_\delta = 25$ MPa-m/s and $\sigma_\delta = 0.83$ Mpa-m/s, respectively.

The press force between the piston shoe and the swash plate is F_P, and the hydraulic reverse thrust is F_w:

$$N = 1.06F_P - F_w$$

$$F_w = \frac{\pi}{2} \frac{r_1^2 - r_2^2}{\ln(r_1/r_2)} p$$

where the parameters in F_w, Figure 3.11, include the slipper sealing outer circle with radius $r_1 = 25.1$ mm, the slipper sealing inner circle with radius $r_2 = 16.25$ mm, and the operational pressure is p.

The contact stress of the slipper is

$$p_d = \frac{N}{A} = p \left[1.06 \frac{\pi d^2}{4 \cos \gamma A} - \frac{1}{2 \ln(r_1/r_2)} \right] = 0.0512p$$

Considering the velocity of piston shoe the stress here is the equivalent stress $s = pv$; therefore, the mean value and standard deviation of the equivalent stress are

$$\mu_s = \frac{p_d v_{\max} + p_d v_{\min}}{2} = 21 \text{ MPa-m/s}$$

$$\sigma_s = \mu_s C = 0.86 \text{ MPa-m/s}$$

- Wear reliability of the piston shoe

Because $Z = \frac{\mu_\delta - \mu_s}{\sqrt{\sigma_\delta^2 + \sigma_s^2}} = 3.367$, the wear reliability of the piston can be obtained from Table A.1 (Appendix A)

$$R_3 = 0.999581 \tag{3.55}$$

4. Wear reliability of the cylinder barrel and valve plate pair

Let's assume that the material of the cylinder barrel surface is CuZn37Mn3AlPbSi and that the material of the valve plate is 38CrMoAl. In this case the mean value and standard deviation of the cylinder barrel equivalent strength are $\mu_\delta = 18.3$ MPa-m/s and $\sigma_\delta = 1.24$ MPa-m/s, respectively. The wear between the cylinder barrel and the valve plate, Figure 3.14, not only depends both on relative velocity and contact force between contact surfaces therefore, the equivalent stress here is selected as $s = pv$.

Similar to the piston shoes, the mean value and standard deviation of the equivalent stress between the cylinder barrel and the valve plate are

$$\mu_s = 11.38 \text{ MPa-m/s}$$

$$\sigma_s = 1.51 \text{ MPa-m/s}$$

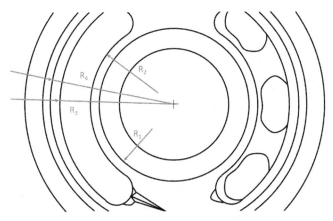

FIGURE 3.14   The valve plate structure.

Here, $Z = \frac{\mu_\delta - \mu_s}{\sqrt{\sigma_\delta^2 + \sigma_s^2}} = 3.54$, and the wear reliability of the cylinder barrel and valve plate can be obtained from Table A.1 (Appendix A)

$$R_4 = 0.9998 \tag{3.56}$$

5. Wear reliability of the piston and cylinder barrel

Let's assume that the material of a piston is 38CrMoAl. In this case its mean value and standard deviation of strength are $\mu_\delta = 18.3$ MPa-m/s and $\sigma_\delta = 1.24$ MPa-m/s, respectively. Figure 3.15 shows the cross-section of the piston. Similar to the previous condition, the equivalent stress is related to the contact force and velocity; that is, $s = pv$. According to the force balance equation of a piston, the mean value and standard deviation of the equivalent stress are

$$\mu_s = 14.1 \ \text{MPa-m/s}$$
$$\sigma_s = 0.665 \ \text{MPa-m/s}$$

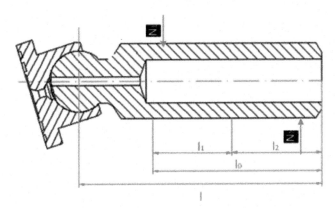

FIGURE 3.15   The stress of a piston.

Here, $Z = \frac{\mu_\delta - \mu_s}{\sqrt{\sigma_\delta^2 + \sigma_s^2}} = 2.985$. The wear reliability of the cylinder barrel and valve plate can be obtained from Table A.1 (Appendix A)

$$R_4 = 0.9986 \tag{3.57}$$

The reliability of a hydraulic pump can be calculated with the series rule as follows

$$R = R_1 R_2 R_3 R_4 = 0.998 \tag{3.58}$$

3.4.3 System Reliability Design

There are two ways of improving reliability: (1) design a component with high reliability or (2) adopt multiple components in a system design. It is becoming increasingly difficult to improve component reliability because design and manufacturing sophistication have resulted in components with near-ultimate reliability. For example, if the failure rate of an integrated circuit needs to be decreased from $\lambda = 5 \times 10^{-7}$ to $\lambda = 1 \times 10^{-7}$ per hour, then it needs to eliminate more than 90% of the components in component screening. Therefore, it is difficult to achieve the reliability requirements simply based on the basis of component reliability improvement. This prompted the development of redundancy techniques to improve system reliability through application of multiple sets of the same components.

3.4.3.1 Definition of Redundancy

According to MIL-F-9490D, redundancy is a design approach such that two or more independent failures, rather than a single failure, are required to produce a given undesirable condition [7]. Redundancy takes the following forms:

1. Providing two or more components, subsystems, or channels, each capable of performing the given function.
2. Monitoring devices to detect failures and automatically disconnect failed components.
3. A combination of the two above features.

Figure 3.16 shows the redundant flight control system and its reliability index appears in Table A.1 (Appendix A).

In Figure 3.16, there are multiple flight control systems, where each channel has triplex sensors and triplex actuators. For a monitoring device set at points A, B, C, D, and E, the corresponding reliability is shown in Table 3.2 [6].

From Table 3.2, the system reliability increases with the increase in redundancy and number of monitoring devices. There are three missions in redundancy design:

1. Determine the fault tolerance capability: Fault tolerance is the property that enables a system to continue operating properly in the event of the failure

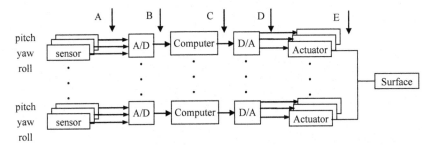

FIGURE 3.16   Redundant flight control system.

of one or more faults within some of its components. The fault-tolerant capability is determined according to the system reliability and safety requirement. The common fault tolerance capabilities include failure operation/failure safe (FO/FS), failure operation/failure operation (FO/FO), and failure operation/failure operation/failure safe (FO/FO/FS). Once the fault tolerance is chosen, the redundancy can be determined.

2. Determine redundancy configuration: Normally, higher redundancy means higher reliability. Figure 3.17 shows that the system reliability increases very fast from two to four components. However, the increase in reliability is diminished when the redundancy number is more than 5. Meanwhile, the failure probability will increase with an increase in the number of components. Therefore, it is necessary to balance the number of redundancies to maintain high reliability and good maintainability. The best redundancy value is currently between 3–5.

3. Determine the redundancy management strategy: Redundancy management is used to ensure that the system is operating to the greatest extent while keeping performance degradation under failure condition at a minimum and providing protection against the fault transient. Redundancy management includes the following:
   a. Signal selection and balance
   b. Failure detection and isolation
   c. System reconfiguration
   d. Resources recovery
   e. Failure detection, recording and displaying

When the redundancy is used in system reliability design, the system reliability can be calculated with a parallel system or a k-out-of N system.

- Parallel system: The parallel system has n components in parallel, and it is considered to be normal if at least one component is good. The RBD of a parallel system is shown in Figure 3.18.

Suppose the life of the ith component is T_i, in which the component is s-independent; then, the life of the parallel system can be described as

$$T_s = \max\{T_1, T_2, \ldots\ldots T_n\} \tag{3.59}$$

TABLE 3.2 The Failure Rate of a Redundant Flight Control System

System Failure Rate	Failure Rate of Single Channel	E	EA	ED	EAD	EB	EC	Triplex Monitoring Voting	
								EBC	EABCD
λ_s per hour	1×10^{-3}	1×10^{-9}	0.78×10^{-9}	0.73×10^{-9}	0.514×10^{-9}	0.435×10^{-9}	0.435×10^{-9}	0.156×10^{-9}	0.134×10^{-9}

Parallel System's Reliability vs. Component Reliability

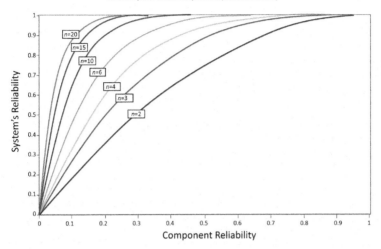

FIGURE 3.17   System reliability as a function of the number of identical components.

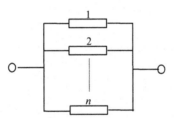

FIGURE 3.18   Reliability diagram of a parallel system.

Then, the system reliability is 8.

$$R_s(t) = P(T_s > t) = P\{\max(T_1, T_2, ..., T_n) > t\}$$
$$= 1 - P\{\max(T_1, T_2, ..., T_n) < t\}$$
$$= 1 - P\{T_1 < t, T_2 < t, ...T_n < t\} \qquad (3.60)$$
$$= 1 - \prod_{i=1}^{n}[1 - R_i(t)]$$

It is apparent that the system reliability in parallel is greater than any of the components' reliability. When the life of the component exhibits the exponential distribution, viz., $R_i(t) = e^{-\lambda_i t}$, $i = 1, 2, ..., n$, the system reliability is

$$R_s(t) = \sum_{i=1}^{n} e^{-\lambda_i t} - \sum_{1 \le i < j \le n} e^{-(\lambda_i + \lambda_j)t} + \sum_{1 \le i < j < k \le n} e^{-(\lambda_i + \lambda_j + \lambda_k)t} + \cdots$$
$$+ (-1)^{n-1} e^{-\left(\sum_{i=1}^{n} \lambda_i\right)t} \qquad (3.61)$$

The *MTTF* of the system is

$$MTTF = \sum_{i=1}^{n} \frac{1}{\lambda_i} - \sum_{1 \le i < j \le n} \frac{1}{\lambda_i + \lambda_j} + \cdots + (-1)^{n-1} \frac{1}{\lambda_1 + \lambda_2 + \cdots + \lambda_n} \quad (3.62)$$

- *k/n(G)* voting system: The system is composed of *n* components. When the system operates under condition when at least *k* components are good, the system is called a *k/n(G)* voting system ($1 \le k \le n$), where *G* means *good*. For instance, the electric power system in an airplane with several engines and cables with many steel wires are considered to be a *k/n(G)* voting system (Figure 3.19).

A *k/n(G)* voting system can have the following three special cases:

1. An *n/n(G)* system is equivalent to a series system with *n* components
2. A *1/n(G)* system is equivalent to a parallel system with *n* components
3. A *m* + 1/(2*m* + 1)(*G*) system is called a *majority voting system*.

Definition:

$$x_i = \begin{cases} 1 & \textit{The ith component is normal} \\ 0 & \textit{The ith component fails} \end{cases}$$

The normal system requires that the following equation must be satisfied:

$$\sum_{i=1}^{n} x_i \ge k \quad (3.63)$$

If the components of the system are identical, then the system reliability can easily be calculated with binomial theorem. Suppose the reliability of a component is *R* and its failure probability is *Q*; then $R + Q = 1$. If the system has *n* components, then the binomial can be expressed as

$$(R + Q)^n = 1$$

After expanding this formula we obtain

$$(R + Q)^n = R^n + C_n^1 R^{n-1} Q + C_n^2 R^{n-2} Q^2 + \cdots + Q^n = 1 \quad (3.64)$$

FIGURE 3.19 *k/n(G)* voting system.

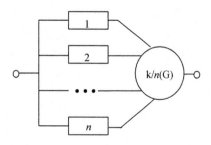

The reliability of the $k/n(G)$ voting system is then

$$R_s(t) = \sum_{i=k}^{n} C_n^i R^i (1-R)^{n-i} \qquad (3.65)$$

If the life of a component has the exponential distribution with parameter λ (i.e., $R_i(t) = e^{-\lambda t}$, $i = 1,2,\ldots,n$), then the reliability of the $k/n(G)$ voting system can be described as

$$R_s(t) = \sum_{i=k}^{n} C_n^i e^{-i\lambda t} \left(1 - e^{-\lambda t}\right)^{n-i} \qquad (3.66)$$

$$MTTF = \int_0^\infty R_s(t)\mathrm{d}t = \sum_{i=k}^{n} C_n^i \int_0^\infty e^{-i\lambda t} \left(1 - e^{-\lambda t}\right)^{n-i}\mathrm{d}t = \sum_{i=k}^{n} \frac{1}{i\lambda} \qquad (3.67)$$

When $k = 1$, the reliability of a $1/n(G)$ system (parallel system) is $R_s = 1 - (1 - R)^n$
When $k = n$, the reliability of an $n/n(G)$ system (series system) is $R_s = R^n$
When $k = m + 1$ and $n = 2m + 1$, the reliability of a majority voting system is

$$R_s(t) = \left[\sum_{i=0}^{m} \binom{2m+1}{i} e^{-\lambda t(2m+1-i)} \left(1 - e^{-\lambda t}\right)^i \right] e^{-\lambda_V t} \qquad (3.68)$$

where λ is the failure rate of a component and λ_V is the failure rate of a voting machine.

Example: Figure 3.20 represents actuation system with three electrical servo units, one mechanical unit, and a cylinder. Calculate the redundant actuation system reliability.

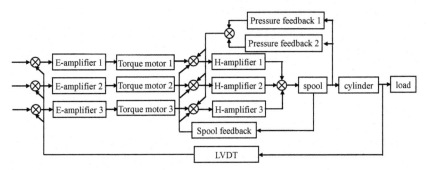

FIGURE 3.20   A kind of actuation system structure.

Solution:

1. E-amplifier reliability: According to the reliability data [7], the failure rate of the electrical amplifier under 0 mA output is $\lambda_{A\_0} = 1 \times 10^{-2}$ per hour, and the failure rate of the electrical amplifier under 10-mA output is $\lambda_{A\_10} = 1 \times 10^{-3}$ per hour. Thus, the electrical amplifier failure probability of the triplex unit is

$$F_{EA} = F_{A\_0}^3 + F_{A\_10}^3 + 3F_{A\_0}\cdot F_{A\_10}^2 + 3F_{A\_0}^2\cdot F_{A\_10}$$
$$= 1 \times 10^{-6} + 1 \times 10^{-9} + 3 \times 10^{-8} + 3 \times 10^{-7}$$
$$= 1.331 \times 10^{-6}$$

The reliability of the triplex electrical amplifier is

$$R_{EA} = 1 - F_{EA} = 0.999998669$$

2. Servo valve reliability: According to the reliability survey [8], the failure rate of a servo valve due to a nozzle clogging is $\lambda_{N\_1} = 1 \times 10^{-2}$ per hour, the failure rate of a spool due to seizing is $\lambda_S = 1 \times 10^{-4}$ per hour, and the failure rate due to a single broken coil is $\lambda_{C\_1} = 1 \times 10^{-2}$ per hour. Thus, the failure probability of the servo valve mechanical structure is as follows:

$$F_{SV-M} = P\{\text{without a pair nozzle}\}$$
$$= 2\cdot F_{N\_1}^3 - F_{N\_1}^6$$
$$= 2 \times 10^{-6} - 1 \times 10^{-12}$$
$$= 1.9999999 \times 10^{-6}$$

The failure probability of three servo coils is

$$F_{SV-C} = F_{C\_1}^3 = 1 \times 10^{-6}$$

Therefore, the reliability of the servo valve is

$$R_{SV} = 1 - (F_{SV-M} + F_S + F_{SV-C} - F_{SV-M}\cdot F_S - F_S\cdot F_{SV-C} - F_{SV-M}\cdot F_{SV-C}$$
$$+ F_{SV-M}\cdot F_S\cdot F_{SV-C})$$
$$= 1 - (1.9999999 \times 10^{-6} + 1 \times 10^{-4} + 1 \times 10^{-6} - 1.9999999 \times 10^{-10}$$
$$- 1 \times 10^{-10} - 1.9999999 \times 10^{-12} + 1.9999999 \times 10^{-16})$$
$$= 0.9998971$$

3. Cylinder reliability: Because the cylinder can be considered as a single-point failure, its reliability is very important in this kind of actuation system. From the reliability report [4], the failure rate of the cylinder due to

seizing is $\lambda_{c\_s} = 1 \times 10^{-5}$ per hour and the failure rate of the cylinder due to cracking is $\lambda_{c\_c} = 1 \times 10^{-5}$ per hour; therefore, the reliability of the cylinder is

$$R_{cy} = 1 - (F_{c\_s} + F_{c\_c} - F_{c\_s} \cdot F_{c\_c})$$
$$= 1 - (1 \times 10^{-5} + 1 \times 10^{-5} - 1 \times 10^{-5} \cdot 1 \times 10^{-5})$$
$$= 0.99998$$

4. LVDT reliability: According to the reliability test, the failure rate of the LVDT is $\lambda_f = 0.5 \times 10^{-4}$ per hour. The reliability of the LVDT is

$$R_f = 1 - (F_f^3 + 3F_f^2(1 - F_f) + 3F_f^2 \cdot F_f)$$
$$= 0.99984731$$

Therefore, the triplex actuation system reliability is

$$R_S = R_{EA}R_{SV}R_{cy}R_f = 0.9997231$$

3.5 DESIGN FOR MAINTAINABILITY

An aircraft hydraulic system consists of a hydraulic power supply system and a hydraulic actuation system. To achieve high reliability and safety of the system a modern hydraulic power supply system adopts dual or triplex hydraulic redundancy. Figure 3.21 shows the hydraulic power supply system in Boeing 757 aircraft [15].

There are three hydraulic power supply systems in Boeing 757 aircraft (left, central, and right hydraulic power supply systems), where each has a reservoir, pump, check valve, safety valve, filter, accumulator, and cooling system. An engine-driven pump (EDP) and AC motor-driven pump (ACMP) are adopted for the left and right hydraulic power supply systems. The central hydraulic power supply system uses two ACMPs. In addition, there are two auxiliary hydraulic systems, a power transfer unit (PTU), and RAT that are used to provide the stand-by power.

Figure 3.22 represents the failure detection system of a single hydraulic power supply system. The system includes an EDP, ACMP, reservoir, check valve, accumulator, filter, safety valve, cooler, and actuator. To maintain high reliability, a single hydraulic power supply system has adopted two pumps (EDP and ACMP). Because the fluid level, temperature, and pressure of a pump are very important parameters, its failure detection system detects the fluid level h and fluid temperature T in the reservoir; output pressure P and temperature T of the EDP and ACMP; and the system pressure.

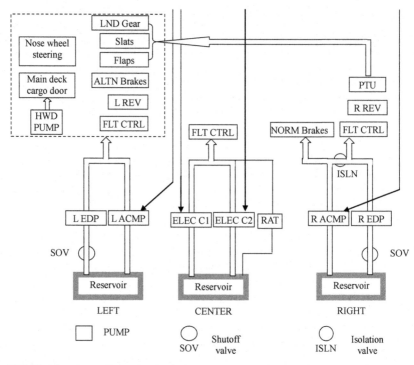

FIGURE 3.21   Hydraulic power supply system of a Boeing 757 [15].

In failure detection system design, it is necessary to consider repair accessibility using the least possible number of sensors. When the check valve, accumulator, or filter fails, the system pressure will change. Therefore, monitoring the system pressure can perform their fault detection.

A redundant hydraulic power supply system is shown in Figure 3.23. The fault detection system focuses on the system pressure and system temperature.

Therefore, the hydraulic indicator system displays the following parameters shown in Figures 3.24–3.26.

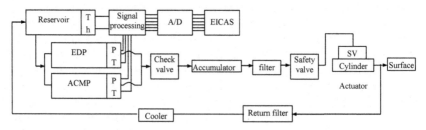

FIGURE 3.22   Failure detection of a single hydraulic power supply system.

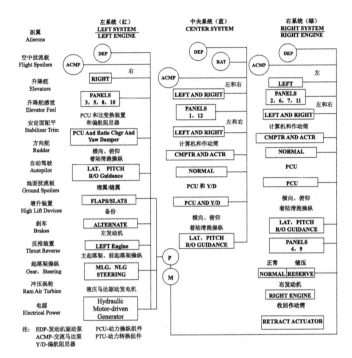

FIGURE 3.23   Failure detection of a redundant hydraulic power supply system [15].

The corresponding failure information is shown in Table 3.3.

Level means the severity of failure, where A indicates the most serious failure that should alarm the pilot, B is the more serious failure that must be ordered in the status page, and C is the normal failure that will be recorded in the status page. M indicates the minor failure that will be entered in the maintenance page, and NVM indicates that the information is stored in RAM.

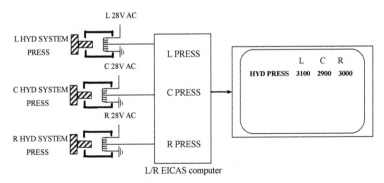

FIGURE 3.24   Hydraulic pressure indicator system [15].

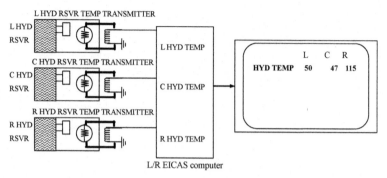

FIGURE 3.25   Hydraulic temperature indicator system [15].

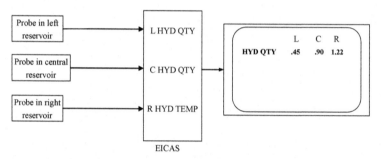

FIGURE 3.26   Hydraulic fluid level indicator system [15].

TABLE 3.3 Failure Information of Hydraulic Power Supply System of Boeing 757 [15]

Information	Level	Failure Detailed
C HYD 1 OVHT	C	ACMP 1 temperature of the central hydraulic system overheat
C HYD 2 OVHT	C	ACMP 2 temperature of the central hydraulic system overheat
C HYD ELEC 1	C	ACMP 1 output pressure low in the central hydraulic system
C HYD ELEC 2	C	ACMP 2 output pressure low in the central hydraulic system
C HYD QTY	C	Fluid level low in the central hydraulic reservoir
C HYD RSVR PRESS	C	Fluid pressure low in the central hydraulic reservoir

TABLE 3.3 Failure Information of Hydraulic Power Supply System of Boeing 757 [15]—cont'd

Information	Level	Failure Detailed
C HYD SYS MAINT	S,M(NVM)	When both engines operate, system pressure low in the central hydraulic system (maintenance page)
C HYD SYS PREES	B	Pressure low in the central hydraulic system
C HYD QTY O/FULL	M	Fluid full in the central hydraulic system
L ENG HYD OVHT*	C	Shell return fluid overheat in the left EDP
L ENG LP PUMP	S,M(NVM)	When engine operates, pressure low in the pump
L HYD ELEC PUMP	C	Pressure low in the left ACMP
L HYD ENG PUMP	C	When engine operates, output pressure low in the left EDP
L HYD QTY	C	Fluid level low in the left hydraulic reservoir
L HYD QTY O/FULL	M	Fluid full in the left hydraulic reservoir
L HYD RSVR PRESS	C	When both engines operate, pressure low in the left hydraulic reservoir
L HYD SYS MAINT	S,M(NVM)	When both engines operate, left hydraulic system pressure is low (maintenance page)
L HYD SYS PRESS	B	Left hydraulic system pressure is low
R GEN DRIVE	C	When engine operates, EDP pressure is low or overheat
R HYD ELEC PUMP	C	ACMP pressure low in the right hydraulic system
R HYD ENG PUMP	C	When engine operates, EDP pressure is low in the right hydraulic system
R HYD QTY	C	Fluid level is low in the right hydraulic reservoir
R HYD QTY O/FULL	M	Fluid full in the right hydraulic reservoir
R HYD RSVR PRESS	C	When both engines operate, fluid pressure is low in the right hydraulic reservoir
R HYD SYS MAINT	S,M(NVM)	When both engines operate, fluid pressure is low in the right hydraulic reservoir (maintenance page)
R HYD SYS PRESS	B	Pressure low in the right hydraulic system

3.6 SAFETY ASSESSMENT METHODS

3.6.1 Failure Modes, Effects, and Criticality Analysis [9]

Failure modes, effects, and criticality analysis (FMECA) is a reliability evaluation/design technique that examines the potential failure modes within a system and its equipment to determine the effects on equipment and system performance. Each potential failure mode is classified according to its effect on mission success and personnel/equipment safety. The FMECA is composed of two separate analyses: the FMEA and the criticality analysis (CA). The FMECA can determine the effects of each failure mode on system performance; provide data for developing FTA and RBD models; provide a basis for identifying root failure causes and developing corrective actions; facilitate investigation of design alternatives to consider high reliability at the conceptual stages of the design; aid in developing test methods and troubleshooting techniques; and provide a foundation for qualitative reliability, maintainability, safety, and logistics analyses.

The FMECA highlights single-point failure requiring corrective action, ranks each failure according to the severity classification of the failure effect on mission success and personnel/equipment safety, estimates system critical failure rates, provides a quantitative ranking of system and/or subsystem failure modes, and identifies reliability/safety critical components.

FMECA benefits include identification of potential design reliability problem areas, information acquisition for troubleshooting activities, maintenance manual development, and design of effective built-in test techniques. The FMECA provides valuable information for maintainability, safety, and logistic analysis.

FMECA is a systematic bottom-up approach to identifying the failure modes of a system, item, or function and determining the effects on the next higher level. FMECAs typically include the following information:

- Identification of component, signal, and/or function
- Failure modes and associated failure rates
- Failure effect (directly and/or at the next higher level)
- Detectability and means of detection
- Severity classification

FMECA may be performed at any level within the system [10]. The FMECA will be completed by identifying the potential failure and cause of failure of each component in the system. The effects of each failure mode are determined by propagating that failure through each level of system structure (local, next higher, assembly, and system level). The failure detection and isolation method and compensating provisions are then recorded.

The CA provides a relative measure of significance of the effect that a failure mode has on the successful operation and safety of the system. CA is

TABLE 3.4 Typical Failure Effect Probabilities β_j

	Actual Loss	Probable Loss	Possible Loss	No Effect
β_{ij}	1.0	0.1–1.0	0–0.1	0

completed after the local, next higher level, and end effects of failure have been evaluated in the FMECA. The value of each failure mode criticality number is defined as

$$C_j = \alpha_j \beta_j \lambda t \tag{3.69}$$

where α_j is the failure mode ratio, β_j is the conditional probability of mission loss, λ is failure rate, and t is the duration of the applicable mission phase given in hours. In general, β_j is quantified in accordance with Table 3.4.

If the system consists of several levels in FMECA, then the criticality summation is

$$C = \sum_{j=1}^{n} C_j = \sum_{j=1}^{n} \alpha_j \beta_j \lambda t \tag{3.70}$$

where n is the number of analytical levels.

After each function at the system level has been analyzed, outputs can be produced and all the information is recorded in Table 3.5.

3.6.2 Fault Tree Analysis [10–12]

FTA is a top-down, deductive failure analysis in which an undesired state of a system is analyzed using Boolean logic to combine a series of lower-level events. This analysis method is mainly used in the fields of safety engineering and reliability engineering to understand how systems can fail, to identify the best ways of reducing risk, or to determine event rates of a safety accident or a particular system level failure. In aircraft, the more general term *system failure* is used for the "undesired state"/top event of the fault tree. These conditions are classified by the severity of their effects. The most severe conditions require the most extensive FTA.

FTA can be used to

- Understand the logic leading to the top event/undesired state
- Show compliance with the system safety/reliability requirements
- Prioritize the contributors leading to the top event
- Create equipment/parts/events lists for different importance measures
- Monitor and control the safety performance of the complex system
- Assist in designing a system.
- Function as a diagnostic tool to identify and correct causes of the top event.

TABLE 3.5 FMECA Worksheet

System _____   Date _____
Part name _____   Sheet _____
Reference drawing _____   Compiled by _____
Mission _____   Approved by _____

Identification Number	Functional Identification	Product Function	Failure Modes and Cause	Mission Phase/ Operational Mode	Failure Effect			Failure Detection Method	Compensating Provisions	Remarks
					Local Effect	Next Higher Level	End Effect			

<table>
<tr><td>Top event</td><td>Basic event</td><td>Undeveloped event</td><td>Conditional event</td><td>Intermediate event</td></tr>
</table>

FIGURE 3.27   Event symbols in FTA.

There are two main components in FTA: events and gates.

The basic symbols used in FTA are grouped as events, gates, and transfer symbols.

3.6.2.1 Event Symbol

The conventional events used in FTA are shown in Figure 3.27.

The primary event symbols are typically used as follows:

- Top event—system failure
- Basic event—failure or error in a system component or element
- Undeveloped event—an event with insufficient information or which is of no consequence
- Conditioning event—conditions that restrict or affect logic gates (e.g., mode of operation in effect)
- Intermediate event—middle event.

3.6.2.2 Logic Gate Symbols

Logic gate symbols describe the relationship between input and output events. The symbols are derived from Boolean logic symbols. There are two types of gates: static gate and dynamic gate, shown in Figure 3.28.

In Figure 3.27, the functions of static logic gates are described below:

- OR gate—the output occurs if any input occurs
- AND gate—the output occurs only if all inputs occur (inputs are independent)
- Exclusive OR gate—the output occurs if exactly one input occurs
- Inhibit gate—the output occurs if the input occurs under an enabling condition specified by a conditioning event

The output of dynamic logic gates occurs if the inputs occur in a specific time sequence specified by the conditioning event.

Transfer symbols are used to connect the inputs and outputs of related fault trees, such as the fault tree of a subsystem to its system.

3.6.2.3 FTA with Static Logic Gate

The tree is usually created by using conventional logic gate symbols shown in Figure 3.29.

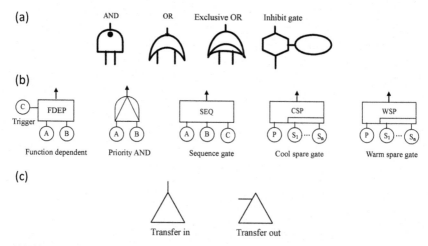

FIGURE 3.28 Conventional logic gate in FTA. (a) Static logic gate, (b) dynamic logic gate, and (c) transfer gate.

The route through a tree between an event and an initiator in the tree is called a *cut set*. The shortest credible way through the tree from fault to initiating event is called a *minimal cut set* (MCS). The most common and popular way to evaluate the FTA can be summarized in a few steps.

FTA analysis involves five steps:

- Define the top event: Definition of the undesired event (top event) can be very hard to identify although some of the events are very easy and obvious to observe. An engineer with a broad knowledge of the system design or a

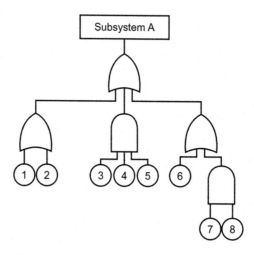

FIGURE 3.29 A typical fault tree diagram.

system analyst with an engineering background is the best person who can help define and number the undesired events.

- Obtain an understanding of the system: Once the undesired event is selected, all causes with probabilities of affecting the undesired event can be obtained. For the selected event, all causes are then numbered and sequenced in the order of occurrence and are used for the next step, which is construction of the fault tree.
- Construct the fault tree: Upon selecting the undesired event and having all of the causal effects, one can construct the fault tree. The fault tree is based on AND and OR gates, which define the major characteristics of the fault tree.
- Evaluate the fault tree: After the fault tree has been assembled for a specific undesired event, one can obtain all of the MCSs [6]. Calculate the probability of MCSs and obtain the failure probability of the system (top event).
- Find the weakness: Using the critical importance [6], we can find the weakness of the system and provide the possible improvement in system design.

3.6.2.4 FTA with Dynamic Logic Gate

Because the dynamic logic gate is related to the time sequence, dynamic fault tree analysis is performed using the Markov analysis.

3.6.3 Markov Analysis

A Markov process is often described by a sequence of directed graphs, in which the state of graph i is labeled by the probabilities of going from one state i to another state j, Figure 3.30.

A Markov process is based on a sequence of random variables $X(t_1)$, $X(t_2)$,... with the Markov property, where the future and past states are independent of the present state. Formally,

$$P\{X(t_n) = i_n | X(t_1) = i_1, ..., X(t_{n-1}) = i_{n-1}\}$$
$$= P\{X(t_n) = i_n | X(t_{n-1}) = i_{n-1}\}$$

(3.71)

where the possible values of $\{X(t), t \geq 0\}$ form a countable set $E = \{i_1, i_2, ... i_n\}$ called *state space* and $0 \leq t_1 < t_2 ... < t_n$ is the moment of state.

FIGURE 3.30   Markov state transfer diagram.

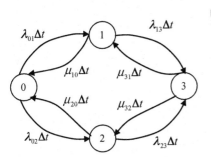

A Markov analysis calculates the probability of the system being in various states as a function of time. A state in the model represents the system status as a function of both the fault-free (state 0 in Figure 3.30) and faulty components and the system redundancy (state 3 in Figure 3.30). A transition from one state to another occurs at a given transition rate (λ is failure rate and μ is reparable rate), based on component failure rates and redundancy. A system changes state due to various events, such as component failure, reconfiguration after detection of failure, completion of repair, etc. Each state transition is a random process that is represented by a specific differential equation [6]. The probability of any state can be determined by solving the associated differential equation. The probability of reaching a defined final state can be determined by combinations of the transitions required to reach that state.

3.6.4 Common Cause Analysis [11]

Independence among functions, systems, or items may be required to satisfy the safety requirements. Therefore, it is necessary to ensure that such independence exists or that the risk associated with dependence is deemed acceptable. Common cause analysis provides the tools to verify this independence or to identify specific dependencies.

Common cause analysis identifies individual failure modes or external events that can lead to a catastrophic or hazardous/severe major failure condition. Such common cause events must be precluded for catastrophic failure conditions and must be within the assigned probability allowance for hazardous/severe major failure conditions.

3.7 COMPREHENSIVE RELIABILITY EVALUATION OF A HYDRAULIC SYSTEM

An aircraft hydraulic system consists of the hydraulic power supply system and hydraulic actuation system; therefore, the comprehensive reliability evaluation focuses on these two systems.

3.7.1 Reliability Evaluation of Hydraulic Power Supply System

Figure 3.31 represents dual-redundant hydraulic power supply systems in A380 aircraft [19], viz. the blue and green system, in which two engines drive four EDPs in each hydraulic system. When the aircraft is on the ground, two electric motor pumps (EMPs) provide the hydraulic power in each hydraulic system. To maintain the safety, two hydraulic power supply systems operate independently.

According to the operational principle, the RBD of the above hydraulic power supply system can be described as shown in Figure 3.32.

The hydraulic power supply systems, Figure 3.32, operate in parallel. The fault tree of a single hydraulic power supply system can be represented as shown in Figure 3.33.

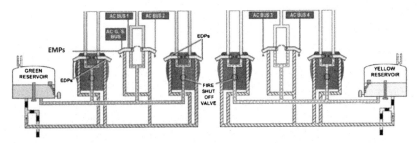

FIGURE 3.31   A380 hydraulic power supply system.

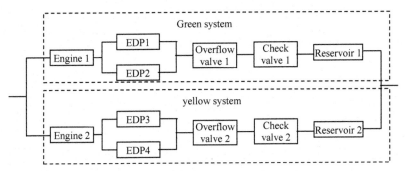

FIGURE 3.32   RBD of dual-redundant hydraulic power supply system.

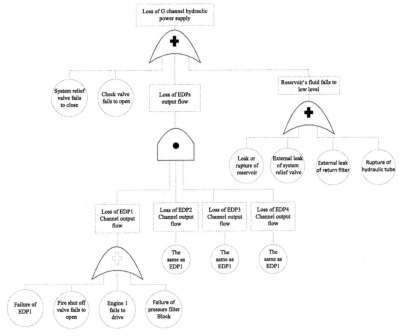

FIGURE 3.33   Fault tree of a single hydraulic power supply system.

TABLE 3.6 Failure Probability of Basic Events

Events	Symbol	Failure Probability
System relief valve fails to close	P_1	$0.2(1 - e^{-26 \times 10^{-6} t})$
Check valve fails to open	P_2	$0.5(1 - e^{-38 \times 10^{-6} t})$
Loss of EDP's output flow	P_3	$1 - e^{-30 \times 10^{-6} t}$
Leak or rupture of reservoir	P_4	$1 - e^{-10 \times 10^{-6} t}$
External leak of system relief valve	P_5	$0.6(1 - e^{-26 \times 10^{-6} t})$
External leak of return filter	P_6	$0.5(1 - e^{-20 \times 10^{-6} t})$
Rupture of hydraulic tube	P_7	$0.4(1 - e^{-10 \times 10^{-6} t})$

In Figure 3.33, symbol ⌂ represents the OR gate and symbol ⌂ represents the AND gate. The failure probability of basic events is presented in Table 3.6 [13].

Applying Boolean algebra, the reliability of a single hydraulic power supply system is

$$R_{hy-1} = 1 - P(A_1 \text{ or } A_2 \text{ or} \ldots A_n)$$

$$= \sum_{i=1}^{n} P(A_i) - \sum_{i\langle j=2}^{n} P(A_i A_j) + (-1)^{n-1} P\left(\bigcap_{i=1}^{n} A_i \right)$$

$$= 0.971$$

According to the reliability theory, the parallel system reliability can be written as

$$R_{hy-S} = 1 - \left(1 - R_{hy-1} \right)\left(1 - R_{hy-2} \right) = 0.999159$$

3.7.2 Safety Requirements and Performance Requirements of the Actuator

Because of the high reliability requirements of a flight control system, several safety functions should be provided by the actuator [14]. Critical surfaces (e.g., aileron, elevator, rudders) are driven by two or three actuators, each powered by an independent hydraulic or electric distribution network. These can be operated in alternating mode (active/standby mode) or in simultaneous mode (active/active mode). Any failure of the neighboring actuators must not lead to a loss of control of the movable surface with the remaining operable actuator.

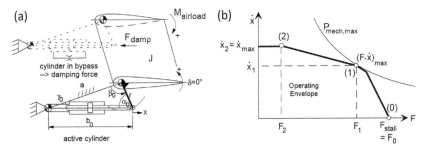

FIGURE 3.34 Redundant actuator structure. (a) Duplex actuator based on HA/EHA, (b) actuator operating point.

To decrease the influence among the actuators of the same surface, some functions such as bypass, movement damping, and load limiting should be implemented. If actuators simultaneously drive the surface, load sensing and compensation should be integrated into the system in order to avoid force fighting.

The effective load/speed operating envelope of the actuator is derived from the load profile of the control surface, taking into account the geometric arrangement shown in Figure 3.34.

The performance requirements include

- The stall load $F_{stall} = F_0$
- The maximum mechanical power $P_{mech,max} = \max(F \cdot \dot{x})$ to be provided by the actuator with the load F_1 (include friction) and speed x_1
- Maximum speed with corresponding load $(\dot{x}_2, F_2)$ or no load

The dynamic performance requirements are shown in Figure 3.35.

3.7.3 Reliability Evaluation of the Hydraulic Actuation System

Modern aircraft use segmentation control surface technology in flight control systems for high reliability. In some critical divided surfaces, a dual-redundant actuation system is also adopted to increase the reliability. Figure 3.36 shows a new kind of application in A380 aircraft that is based on a hydraulic actuator (HA) and electrohydraulic actuator (EHA), in which HA operates actively and EHA follows under normal operating conditions. The EHA assumes operation in the case of HA failure.

In Figure 3.36, the HA section consists of the hydraulic power supply system, servo valve, cylinder, and LVDT. The input of the servo valve is i_v, the output force of the HA is F_h, and the displacement of the cylinder is x_h. The EHA section consists of a brushless motor, pump, and cylinder, in which the input to the motor is the control voltage u_e, the output force is F_e, and the displacement of the cylinder is x_e. The inputs to the control surface are the

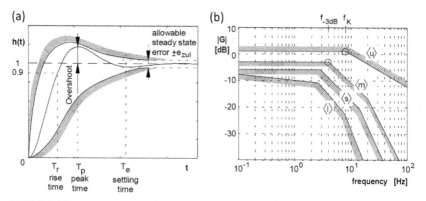

FIGURE 3.35 Dynamic performance requirement of an actuator. (a) Step response of actuator, (b) frequency response of actuator.

displacement of x_h, x_e, and aerodynamic load F_L, whereas the outputs are the surface displacement x_t and the force acted in both cylinders, F_h and F_e.

In the case when HA is active and EHA is a follower, mathematical model is developed as indicated below.

1. Servo valve model

Suppose the servo valve control current is i_v, the servo valve spool displacement is x_v, and the gain of the amplifier is K_v. The transfer function of the servo valve can then be described as a second-order system:

$$Q_h = \frac{K_q K_v \omega_v^2}{s^2 + 2\xi_v \omega_v s + \omega_v^2} i_v - K_c p_h \tag{3.72}$$

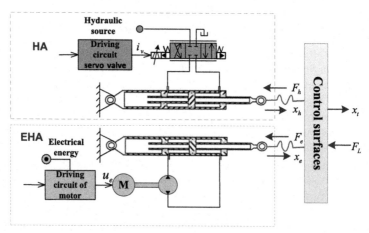

FIGURE 3.36 Dual-redundant actuation system based on HA and EHA.

where ω_v is the characteristic frequency of the servo valve, ξ_v is the damping coefficient of the servo valve, Q_h is the load flow, p_h is the load pressure, k_q is the flow gain, and K_c is the flow-pressure coefficient.

2. Cylinder model

Suppose that the fluid has laminar flow, that the fluid temperature is constant, and that the friction loss and pipe dynamic influence can be neglected. The flow equation of the cylinder can then be written as

$$Q_h = A_h \frac{dx_h}{dt} + \frac{V_{th}}{4E_y} \frac{dp_h}{dt} + C_{sh}p_h \qquad (3.73)$$

The force balance equation of the cylinder is

$$A_h p_h = m_{ph}\frac{d^2 x_h}{dt^2} + B_{ph}\frac{dx_h}{dt} + F_h + f_h \qquad (3.74)$$

where A_h is the piston area, x_h is the displacement of HA, V_{th} is the total volume of HA, E_y is the equivalent volume elastic modulus, C_{sh} is the total leakage coefficient, m_{ph} is the piston mass, B_{ph} is the viscous damping coefficient, and f_h is the friction between the piston and the cylinder.

3. HA/EHA integrated model

HA adopts the proportional control, and EHA connects the two chambers to follow the HA as a damping force f_e. Considering that the equivalent mass and connection stiffness of a surface are m_d and K_t, respectively, the motion equation of the control surface can be written as

$$m_d \frac{d^2 x_t}{dt^2} = F_h - F_e - F_L \qquad (3.75)$$

$$F_h = K_t(x_h - x_t) \qquad (3.76)$$

$$F_e = K_t(x_t - x_e) \qquad (3.77)$$

$$F_e = m_{pe}\frac{d^2 x_e}{dt^2} + B_{pe}\frac{dx_e}{dt} \qquad (3.78)$$

where m_{pe} is the mass of EHA and B_{pe} is the viscous damping coefficient of EHA.

Taking the Laplace transform of the Equations 3.75–3.78, the control block diagram of the HA/EHA actuator is represented in Figure 3.37.

When the cylinder operates, the wear and leakage will increase, Figure 3.38. The leakage flow can be expressed as

$$Q = \frac{\pi d \delta^3 \Delta P}{12\mu LC} \qquad (3.79)$$

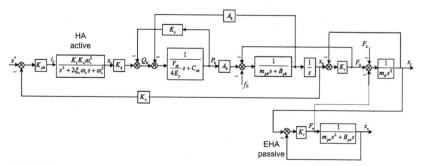

FIGURE 3.37 System block diagram under HA active, EHA follower.

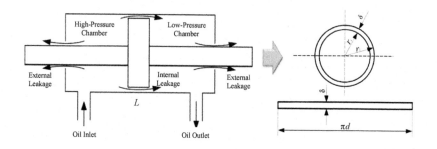

FIGURE 3.38 The leakage diagram of a cylinder.

The leakage coefficient can be calculated as

$$C_{sh} = \frac{Q}{\Delta P} = \frac{\pi d \delta^3}{12 \mu L C} \tag{3.80}$$

where d is the piston diameter, δ is the single slit height, ΔP is the pressure difference between two chambers, μ is the fluid dynamic viscosity, L is the travel distance of the piston, and C is the initial laminar correction coefficient.

With the operation of the cylinder, the wear volume between the cylinder and the piston can be described with the abrasive wear formula [13]

$$\Delta V = K_s \frac{WL'}{H} \tag{3.81}$$

where ΔV is the wear volume, W is the normal load, H is the material hardness, K_s is the abrasive wear coefficient, and L' is the sliding distance.

Considering the number of actuation cycles n, the wear volume can be described as

$$\Delta V = K_s \frac{W n l_0 t}{H} \tag{3.82}$$

After replacing the wear volume into the leakage coefficient relationship, we obtain

$$C_{sh} = \frac{\pi d(\delta_0 + \Delta\delta)^3}{12\mu LC} = \frac{\pi d\left(\delta_0 + \frac{K_s Wnl_0}{H\pi dL}t\right)^3}{12\mu LC} \tag{3.83}$$

where δ_0 is the initial slit height.

Considering the external load disturbance and wear unevenness, the leakage coefficient is subject to a normal distribution as follows

$$f_{C_{sh}}(C_{sh}, t) = \frac{1}{\sqrt{2\pi}\sigma}\exp\left\{-\frac{\left[C_{sh} - \frac{\pi d\left(\delta_0 + \frac{K_s Wnl_0}{H\pi dL}t\right)^3}{12\mu LC}\right]^2}{2\sigma^2}\right\} \tag{3.84}$$

The performance degradation curve can be obtained for different leakage coefficient, Figure 3.39.

4. System performance reliability evaluation

Considering the performance degradation due to leakage, define the performance reliability as

$$R_Y(t) = P\{\mathbf{Y} \in \mathbf{\Omega}|\mathbf{X} \sim \mathbf{D}\} \tag{3.85}$$

where $\mathbf{Y}$ is the system performance, $\mathbf{\Omega}$ is the performance threshold, $X = \{X_1, X_2, \ldots, X_n\}$ is the system parameter set, D is the probability distribution of

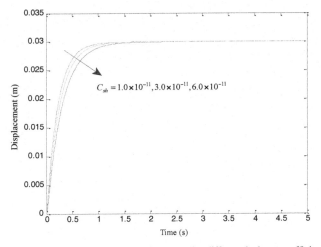

FIGURE 3.39 The dynamic performance response under different leakage coefficients.

the system parameters, and $Y \in \Omega$ is the event that the performance meets the requirement.

Because the aircraft actuation system requires a very rapid response, the response time needs to be selected to describe the performance reliability as

$$R_Y(t) = P\left\{T_r < v_{T_r} \middle| K_j \sim f_{K_j}(k_j, t)\right\} \tag{3.86}$$

where v_{T_r} is the threshold of the response time, $K_j = \{K_q^j, K_v^j, K_c^j, C_{sh}^j\}$ is the system parameter set, and $f_{K_j}(k_j, t)$ is the probability density of the system parameters.

According to the mathematical model in Figure 3.35, the system transfer function is

$$G(s) = \frac{A_h K_Q(s) K_t K_{ph}}{H(s) + A_h K_Q(s) K_{ph} G_{xF}(s) + G_{a4}(s) G_{xF}(s)} \tag{3.87}$$

Herein,

$$K_Q(s) = K_q \frac{K_v \omega_v^2}{s^2 + 2\xi_v \omega_v s + \omega_v^2} \tag{3.88}$$

$$
\begin{aligned}
H(s) = {} & \frac{V_{th} m_{ph} m_d}{4E_y} s^5 + \left(\frac{V_{th} m_d B_{ph}}{4E_y} + K_{tm} m_{ph} m_d\right) s^4 + \left(K_{tm} B_{ph} m_d + A_h^2 m_d\right. \\
& + \frac{V_{th} K_t m_d}{4E_y} + \frac{V_{th} m_{ph} K_t}{4E_y}\right) s^3 + \left(\frac{V_{th} K_t B_{ph}}{4E_y} + K_{tm} K_t m_{ph} + K_{tm} K_t m_d\right. \\
& + A_h K_Q(s) K_{ph} m_d\right) s^2 + \left(K_{tm} B_{ph} K_t + K_t A_h^2\right) s + A_h K_Q(s) K_{ph} K_t
\end{aligned}
$$
$$\tag{3.89}$$

$$
\begin{aligned}
G_{a4}(s) = {} & \frac{V_{th}}{4E_y} m_{ph} s^3 + \left(\frac{V_{th}}{4E_y} B_{ph} + K_{tm} m_{ph}\right) s^2 \\
& + \left(K_{tm} B_{ph} + A_h^2 + \frac{V_{th}}{4E_y} K_t\right) s + K_{tm} K_t
\end{aligned}
\tag{3.90}
$$

$$G_{xF}(s) = \frac{K_t\left(m_{pe} s^2 + B_{pe} s\right)}{\left(m_{pe} s^2 + B_{pe} s + K_t\right)} \tag{3.91}$$

$$K_{tm} = K_c + C_{sh} \tag{3.92}$$

With the aforementioned transfer function, the system performance can be described as

$$T_r = G_{T_r}(K_j) \tag{3.93}$$

The parameter distribution is

$$K_j \sim f_{K_j}(k_j, t) \tag{3.94}$$

The performance distribution can then be described as

$$T_r \sim g_{T_r}(t_r, t) \tag{3.95}$$

The performance reliability of actuation system can then be expressed as

$$P = \int_t^\infty \int_0^{v_{T_r}} g_{T_r}(t_r, t)\mathrm{d}t_r\mathrm{d}t \tag{3.96}$$

Because the mathematical model is very complicated, it is difficult to determine analytical solution to the performance reliability; therefore, one needs to resort to a numerical solution. The simulation flow chart shown in Figure 3.40 can result in a solution to the performance reliability of the actuation system.

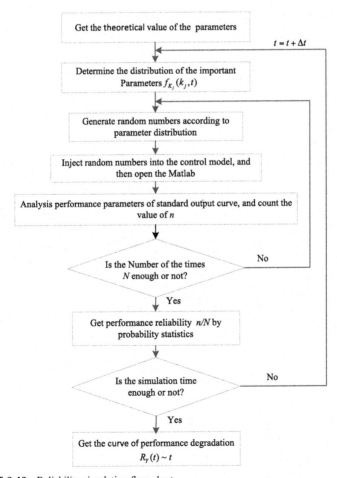

FIGURE 3.40 Reliability simulation flow chart.

TABLE 3.7 The Component Failure Rate

Number	Component	Failure Rate (per hour)
1	Amplifier of servo valve	300×10^{-6}
2	Servo valve	360×10^{-6}
3	Cylinder	270×10^{-6}
4	LVDT	150×10^{-6}
5	Bypass valve	200×10^{-6}

The reliability of the HA/EHA system depends not only on the performance degradation, but it can also be related to the critical component reliability. Assuming that the performance reliability is independent of the component reliability, we can define the integrated reliability of the actuation system as follows:

$$R(t,y) = P\{\mathbf{Y} \in \mathbf{\Omega}, T > t\}$$
$$= P\{\mathbf{Y} \in \mathbf{\Omega}|T > t\} \cdot P\{T > t\} \tag{3.97}$$
$$R(t,y) = R_T(t) \cdot R_Y(t)$$

where the $R_T(t)$ is the component reliability. Because the main components of the actuation system consist of an amplifier, servo valve, cylinder, LVDT, and bypass valve, the component reliability is as follows:

$$R_T(t) = \prod_{i=1}^{5} R_T^i(t) = \prod_{i=1}^{5} e^{-\lambda_i t} \tag{3.98}$$

where $R_T^i(t)$ is the reliability of components and λ_i is the component failure rate. The failure rate of individual components is listed in Table 3.7.

TABLE 3.8 System Parameter Value

Parameter	Value	Unit	Parameter	Value	Unit
ζ_v	0.7	–	V_{th}	1.47×10^{-4}	m^3
ω_v	600	rad/s	E_y	8.0×10^8	Pa
K_v	1.52×10^{-4}	m/A	C_{sh}	1.0×10^{-11}	$(m^3/s)/Pa$
K_q	2.7	m^2/s	K_t	1×10^8	N/m
K_c	1.75×10^{-11}	$(m^3/s)/Pa$	m_d	600	Kg
$m_{ph}(m_{pe})$	55	kg	$B_{ph}(B_{pe})$	10,000	Ns/m

(a)
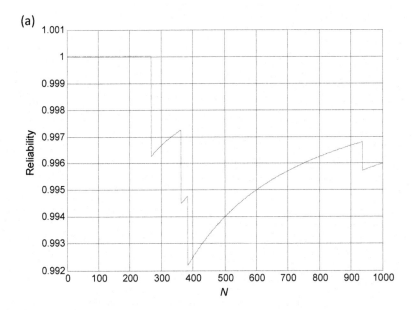

(b)
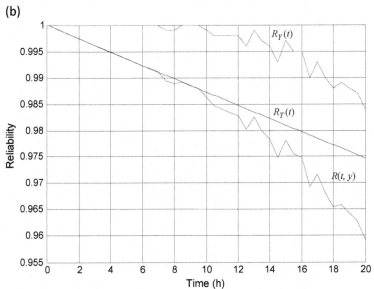

FIGURE 3.41   Integrated reliability of an actuation system. (a) Performance reliability curve at $t = 14$ h, (b) integrated reliability curve under function and performance reliability.

The reliability of the component is then $R_T(t) = e^{-0.00128t}$.

The required parameters to calculate the performance reliability are listed in Table 3.8.

Assuming the simulation number $N = 2000$, the time interval as $\Delta t = 0.5$ h, the response time $T_r < 0.6$ s, and the step input $x^* = 0.03$ m, and substituting the parameters from Table 3.8 into the performance reliability, we obtain the performance reliability curve shown in Figure 3.41.

It is apparent that the integrated reliability takes into consideration the performance degradation with the system operation in addition to component reliability. Furthermore, increasing the simulation number improves the accuracy of the result and makes it closer to the real application.

3.8 CONCLUSIONS

This chapter introduces the reliability design methods for components and systems. The mechanical component reliability design depends on the stress-strength interference theory, and the system reliability design is related to redundancy and monitoring. In addition, this chapter provides the maintenance design theory for hydraulic systems. Through allocation of the monitoring sensors, the corresponding fault could be automatically detected, diagnosed, and isolated. Furthermore, this chapter provides the performance reliability definition and the reliability calculation methods. Afterward, the integrated reliability evaluation is provided for example of the hydraulic power supply system and the actuation system. The results indicate that the system design meets the desired system reliability requirements.

REFERENCES

[1] MIL-F-9490E, Flight Control Systems—Design, Installation and Test of Piloted Aircraft, Central Specification for, 2008.

[2] D. Pettit, A. Turnbull, General Aviation Aircraft Reliability Study, NASA/CR-20110—210647, FDC/NYMA Inc., Hampton, Virginia, February 2001.

[3] J. Mclinn, A Short History of Reliability, http://kscsma.ksc.nasa.gov/Reliability/Documents/History_of_Reliability.pdf.

[4] Military Standard-Definition of Terms for Reliability and Maintainability, MIL-STD-721B, 1966.

[5] Reliability of Military Electronic Equipment Report, US Govt. Print. Off., Washington, 1957 (HATHI TRUST Digital Library).

[6] S. Wang, Reliability Engineering, Beijing University of Aeronautics and Astronautics Press, 2001 (in Chinese).

[7] Reliability Handbook of Electric Components, GJB/Z 299C, China, 2006.

[8] Reliability Handbook of Non-electric Components, GJB/Z 108A, 2006.

[9] R. Borgovini, S. Pemberton, M. Rossi, Failure Mode, Effects and Criticality (FMECA), AD-A278 508, Reliability Analysis Center, PO Box 4700, Rome, NY 13442-4700, 1993.

[10] Fault Tree Analysis. http://en.wikipedia.org/wiki/Fault_tree_analysis.

[11] Aerospace Recommended Practice, SAE International, ARP4761, 1996.

[12] Center for Chemical Process Safety, Guidelines for Hazard Evaluation Procedures, third ed., Wiley, ISBN 978-0-471-97815-2, 2008.

[13] Q. Li, Reliability Model and Evaluation for Aircraft Redundant Power Supply and Actuation System, Master Degree Thesis of Beihang University, 2014.

[14] S. Frischemeier, Electro-hydrostatic Actuator for Aircraft Primary Flight Control Types, Modeling and Evaluation, The 5th Scandinavian International Conference, Frischemeier, pp 1−16, 1997.

[15] Boeing Airplane Maintenance Manual, Seattle, Boeing, 2000.

[16] E.A. Elsayed, Reliability Engineering, Wiley, 2011.

[17] RADC-TR-68-403, in: C. Lipson, et al. (Eds.), Reliability Prediction—Mechanical Stress/Strength Interference (Nonferrous), Rome Air Development Center, Air Force Systems Command, Griffiss Air Force Base, New York, February 1969.

[18] RADC-TR-66-710, in: C. Lipson, et al. (Eds.), Reliability Prediction—Mechanical Stress/Strength Interference, Rome Air Development Center, Air Force Systems Command, Griffiss Air Force Base, New York, March 1967.

[19] Peter A. Stricker, Aircraft hydraulic system design, Eaton Aerospace Hydraulic System Division Report, 2010.

Chapter 4

New Technology of Aircraft Hydraulic System

Chapter Outline

4.1 INTRODUCTION

With the rapid growth of air travel, it is necessary to consider aircraft carbon emissions. In the European Union, the greenhouse gas emissions from aviation increased by 87% from 1990 to 2006. According to data from the United Nations Intergovernmental Panel on Climate Change, aviation produced around 2% of the world's manmade emissions of carbon dioxide in 2011. As aviation grows to meet increasing demands, the Intergovernmental Panel on Climate Change forecasts that its share of global manmade carbon dioxide emissions will increase to around 3% by 2050. However, despite growth in passenger numbers at an average rate of 5% each year, aviation has managed to decouple its emissions growth by around 3% (or some 20 million tons annually). This is achieved through massive investment in new technologies and operating procedures. As the world's two biggest aircraft manufacturers, Airbus and Boeing have made great efforts to improve their best-selling single-aisle aircrafts A320 and B737 to meet this objective. Their new models under development are A320Neo and B737Max, will have improved fuel efficiency compared with the current models. The aviation industry is making continuous efforts to develop the next generation of aircraft which will be increasingly greener, cheaper, and safer with an increased passenger appeal. Trends in the development of next-generation aircraft can be summarized as shown in the following figure [1].

Commercial Aircraft Hydraulic Systems. http://dx.doi.org/10.1016/B978-0-12-419972-9.00004-8
171

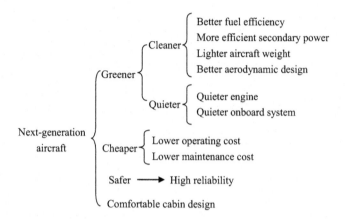

One of the areas where improvements can be made includes the hydraulic system. A hydraulic power supply system provides high-pressure fluid for ailerons, elevators, rudders, landing gear, flaps, spoilers, etc. With the continuous increase in aircraft size, velocity, and overall maneuverability, aircraft require a lighter flying platform and ability to handle larger payload. As the main secondary power of current aircraft, the conventional hydraulic power supply system has heavy pipes between the pump and the actuation system, and has low efficiency in the throttle regulation and leakage. To decrease the weight of centralized hydraulic power supply systems, there are some new technologies in aircraft design:

1. High-pressure hydraulic power supply. Increasing the fluid pressure results in a lighter structure.
2. Strategy for adjustment of the system pressure and flow according to the flight profile to save energy consumption.
3. Application of electrical power to replace the hydraulic network in order to achieve high reliability.

This chapter introduces some of the new technologies, including:

1. High-pressure hydraulic power suply system to reduce weight and volume.
2. An intelligent variable pressure hydraulic system to save energy.
3. More applications of electrical components and systems to decrease the number of pipes between the centralized hydraulic power system and the actuation system.

4.2 HIGH-PRESSURE, HIGH-POWER HYDRAULIC AIRCRAFT POWER SUPPLY SYSTEMS

4.2.1 Introduction of High Pressure

Since the advent of aircraft hydraulic power supply, its pressure has been kept at 3000 psi (21 MPa) for more than 40 years [2]. Figure 4.1 shows the operational pressure used in different types of aircraft.

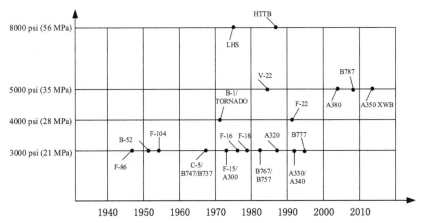

FIGURE 4.1 Operational pressure in different kinds of aircraft.

In 1938, the Douglas Corporation first adopted the hydraulic power supply system in DC-4 aircraft. In the 1940s, the rated pressure of aircraft hydraulic system was between 500 psi (3.5 MPa) and 1000 psi (7 MPa) [3]. The corresponding pump was the gear pump, in which the relief valve was used to adjust the pump pressure. In the 1950s, Martin proposed to the Air Force that 3000 psi (21 MPa) pressure was a practical upper limit of aircraft hydraulic systems, which has been in effect ever since [4]. Aircraft hydraulic systems of this period use a constant pressure variable pump as the power supply source. Although aircraft hydraulic system operational pressure did not improve over the years, its power increased from 22.4 to 298.3 KW [5]. In the mid-1970s, the United States and some European countries began to use a 4000-psi (35 MPa) hydraulic system, installing it on a number of models, such as the Mirage 2000, B-1B, "Concord," etc. [6]. In the early 1980s, the US Air Force and Navy worked on development of high-pressure hydraulic systems. The newly developed aircraft, such as JAS39, F-22, B-2, and C-17, adopted 4000 psi (28 MPa) use [7]. Russia's Sukhoi aircraft design bureau also developed the "SU" series aircraft with a 4000-psi (28 MPa) hydraulic system [8]. In the early 1990s, the newly developed "Rafale" fighter jets and V-22 aircraft were equipped with a 5000 psi (35 MPa) system, which is still the highest rated operational pressure hydraulic systems in the world today [9]. At the end of the twentieth century, the B787 and A380 improved their hydraulic pressure to 5000 psi (35 MPa) [10]. Rockwell did some simulations and experiments to prove that the optimal fluid pressure of aircraft hydraulic system is 8000 psi (56 MPa) [11]. Studies have shown that the A380 was able to reduce its overall weight by 1.8 T after adopting a 5000-psi hydraulic system. Comparative studies showed that by increasing hydraulic pressure from 3000 psi (21 MPa) to

8000 psi (56 MPa), the weight of a hydraulic system can be reduced by 30%, whereas its volume can be reduced by 40%.

With the increasing demand for high performance, there are some new characteristics in next-generation aircraft design:

1. High-Mach cruise requires small and thin airfoil which limits the installation space of hydraulic actuator, and demands smaller volume of the hydraulic actuator. The US Air Force requires the installation of hydraulic components inside the wing volume reduction of 60% [12].
2. Increasingly efficient use of engine power to improve the thrust-to-weight ratio, reduce weight, and improve the flight performance. The weight of a current hydraulic power supply system is 3—15% of the weight of the whole aircraft, whereas the desired weight of hydraulic system is 1% of the weight of the whole aircraft [13].
3. Supersonic aircraft mobility can withstand greater aerodynamic loads for moving at a faster rate, but the hydraulic system requires more power. YF-22A fighter aircraft hydraulic systems' power is close to 600 kW, about twice that of F-15 fighter aircraft [14], which has led to increased volume and weight of the hydraulic system.
4. High-reliability application of the aircraft hydraulic system requires implementation of a redundancy technique [15—17], resulting in an increased number of hydraulic system components and thus increased volume and weight of the installation.

These requirements on weight, volume, power, payload, performance, and reliability impose conflicting demands on the design of aircraft hydraulic system [18]. Thus, an increase in hydraulic system pressure is the most effective way to reduce the weight and volume of hydraulic system [16]. Table 4.1 shows the relation between fluid pressure and system weight and volume reduction [19].

TABLE 4.1 Hydraulic System Weight Reduction and Fluid Operational Pressure

The Situation of Improving Pressure		3000—4000 psi (21—28 MPa)	3000—5000 psi (21—35 MPa)	3000—8000 psi (21—56 MPa)
System changes	Weight reduction	2.46%	12.2%	30%
	Volume reduction	13.8%	28.3%	40%

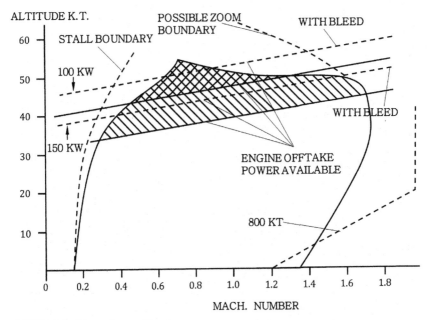

FIGURE 4.2   The engine provides the power.

4.2.2 High Power

The future design of aircraft will continue to push the envelope of speed and maneuverability, which will require reduction in the hydraulic system weight and volume as well as increase in power. For example, the power of the Su27 is eight times that of Chinese aircraft designed in the past century, which makes the Su27 more maneuverable [8]. In a high-altitude flight, aircraft need maximum power some times. In this situation, the power that the engine provides to the hydraulic system is not sufficient even with the accumulator shown in Figure 4.2.

The anticipated required hydraulic power of the future high-performance aircraft is almost five times that of the present aircraft [6]. Hence, the trend of future aircraft hydraulic systems development is toward increasing power. For example, the power of an F-22 reaches 560 KW; the power of an A380 approaches 600 KW [14]. As a result, the very important issue for hydraulic power supply systems is maintaining high efficiency under high pressure and high power.

4.2.3 The Problem of High-Pressure Hydraulic Systems

Although a high-pressure hydraulic system can decrease the weight and volume of hydraulic components, it also brings new problems, as discussed in the following paragraphs.

4.2.3.1 Noneffective Power Increases Caused by High Pressure

In the hydraulic system, the flow loss of a hydraulic pump under constant temperature can be calculated by:

$$Q_1 = K_1 p_s \tag{4.1}$$

where, K_1 is the leakage coefficient of the pump (approximately constant, and specific system-related), $m^5/(N \cdot s)$; p_s is the pump supply pressure, Pa.

The power loss, P_1, due to the volume loss , Q_1 , is

$$P_1 = Q_1 p_s \tag{4.2}$$

Substituting Eqn (4.2) into Eqn (4.1) results in

$$P_1 = K_1 p_s^2 \tag{4.3}$$

The power loss of pump is proportional to the square of pressure, which means that increasing the fluid pressure will result in increased leakage in the hydraulic system. There is a similar law of the leakage for other hydraulic components. For example, as the pressure of a hydraulic pump increases from 1500 psi (10.5 MPa) to 3000 psi (21 Mpa), the volume of power loss increases from 0.735 to 2.94 KW [15].

4.2.3.2 Noneffective Power Increase Caused by High Power

With the increasing of hydraulic system power, the power losses of the hydraulic system also increase. Under high-power conditions, the power losses increase because the pump cannot match the load. When the load requires low pressure, most of the hydraulic power is consumed in the throttling loss. According to the relationship between the hydraulic pump and the actuation system, Figure 4.3, the shaded area indicates power loss in the system.

Aircraft hydraulic system have developed to higher pressure and higher power, which leads to an increase in power loss, which results in excessive heat generation and fluid temperature. An increase in temperature accelerates the aging process in fluid and seals. An expected increase in aircraft hydraulic system pressure from 3000 psi (21 MPa) to 8000 psi (56 MPa) will result in a temperature increase from 110 to 180 °C. The statistic data indicate that the medium stable life will reduce 90% whenever the fluid temperature rises by 15 °C. An extremely adverse impact of the fluid temperature lies in reducing the viscosity and lubricating properties, which brings certain issues related to sealing difficulties in high-pressure hydraulic systems. In addition, an increase in temperature intensifies the sediment aggregation, reduces lubrication, and can result in the system failure [20]. Moreover, future aircraft will increasingly use composite materials that are characterized by poor heat transmission capacity. If the aircraft flies at supersonic speeds, an increase in the aircraft shell temperature will further increase the hydraulic system temperature. Hence, it

After replacing the wear volume into the leakage coefficient relationship, we obtain

$$C_{sh} = \frac{\pi d(\delta_0 + \Delta\delta)^3}{12\mu LC} = \frac{\pi d\left(\delta_0 + \frac{K_s Wnl_0}{H\pi dL}t\right)^3}{12\mu LC} \tag{3.83}$$

where δ_0 is the initial slit height.

Considering the external load disturbance and wear unevenness, the leakage coefficient is subject to a normal distribution as follows

$$f_{C_{sh}}(C_{sh},t) = \frac{1}{\sqrt{2\pi}\sigma}\exp\left\{-\frac{\left[C_{sh} - \frac{\pi d\left(\delta_0 + \frac{K_s Wnl_0}{H\pi dL}t\right)^3}{12\mu LC}\right]^2}{2\sigma^2}\right\} \tag{3.84}$$

The performance degradation curve can be obtained for different leakage coefficient, Figure 3.39.

4. System performance reliability evaluation

Considering the performance degradation due to leakage, define the performance reliability as

$$R_Y(t) = P\{\mathbf{Y} \in \mathbf{\Omega}|\mathbf{X} \sim \mathbf{D}\} \tag{3.85}$$

where $\mathbf{Y}$ is the system performance, $\mathbf{\Omega}$ is the performance threshold, $X = \{X_1, X_2, \ldots, X_n\}$ is the system parameter set, D is the probability distribution of

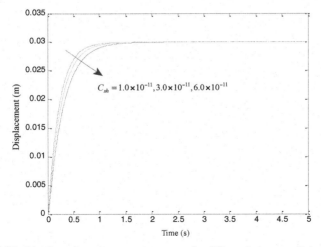

FIGURE 3.39 The dynamic performance response under different leakage coefficients.

the system parameters, and $Y \in \Omega$ is the event that the performance meets the requirement.

Because the aircraft actuation system requires a very rapid response, the response time needs to be selected to describe the performance reliability as

$$R_Y(t) = P\left\{T_r < v_{T_r} \middle| K_j \sim f_{K_j}(k_j, t)\right\} \tag{3.86}$$

where v_{T_r} is the threshold of the response time, $K_j = \{K_q^j, K_v^j, K_c^j, C_{sh}^j\}$ is the system parameter set, and $f_{K_j}(k_j, t)$ is the probability density of the system parameters.

According to the mathematical model in Figure 3.35, the system transfer function is

$$G(s) = \frac{A_h K_Q(s) K_t K_{ph}}{H(s) + A_h K_Q(s) K_{ph} G_{xF}(s) + G_{a4}(s) G_{xF}(s)} \tag{3.87}$$

Herein,

$$K_Q(s) = K_q \frac{K_v \omega_v^2}{s^2 + 2\xi_v \omega_v s + \omega_v^2} \tag{3.88}$$

$$H(s) = \frac{V_{th} m_{ph} m_d}{4E_y} s^5 + \left(\frac{V_{th} m_d B_{ph}}{4E_y} + K_{tm} m_{ph} m_d\right) s^4 + \left(K_{tm} B_{ph} m_d + A_h^2 m_d\right.$$
$$+ \frac{V_{th} K_t m_d}{4E_y} + \frac{V_{th} m_{ph} K_t}{4E_y}\right) s^3 + \left(\frac{V_{th} K_t B_{ph}}{4E_y} + K_{tm} K_t m_{ph} + K_{tm} K_t m_d\right.$$
$$+ A_h K_Q(s) K_{ph} m_d\right) s^2 + \left(K_{tm} B_{ph} K_t + K_t A_h^2\right) s + A_h K_Q(s) K_{ph} K_t \tag{3.89}$$

$$G_{a4}(s) = \frac{V_{th}}{4E_y} m_{ph} s^3 + \left(\frac{V_{th}}{4E_y} B_{ph} + K_{tm} m_{ph}\right) s^2$$
$$+ \left(K_{tm} B_{ph} + A_h^2 + \frac{V_{th}}{4E_y} K_t\right) s + K_{tm} K_t \tag{3.90}$$

$$G_{xF}(s) = \frac{K_t \left(m_{pe} s^2 + B_{pe} s\right)}{\left(m_{pe} s^2 + B_{pe} s + K_t\right)} \tag{3.91}$$

$$K_{tm} = K_c + C_{sh} \tag{3.92}$$

With the aforementioned transfer function, the system performance can be described as

$$T_r = G_{T_r}(K_j) \tag{3.93}$$

The parameter distribution is

$$K_j \sim f_{K_j}(k_j, t) \tag{3.94}$$

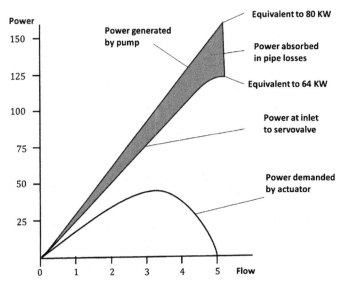

FIGURE 4.3   The power relation between the hydraulic pump and the actuation system.

is essential to find a way to manage the energy, decrease the power loss, and improve the hydraulic system efficiency.

Because hydraulic power is produced by a hydraulic pump, energy management techniques should also focus on the hydraulic power supply. There are three options that can be considered for hydraulic energy-saving:

1. application of an accumulator
2. application of a constant pressure variable pump
3. application of a load-sensing pump and power matching circuit.

The early hydraulic power supply systems consisted of constant rate pump with relief valve, as shown in Figure 4.4.

The pump outlet pressure, Figure 4.4, is determined by the relief valve. When the load flow demand is close or equal to the output flow of constant rate pump, the energy efficiency is highest. In real applications, part or all of pump flow will return to the reservoir through a relief valve, so the efficiency of this kind of hydraulic power supply is too low to be used.

The next development in hydraulic power supply systems was the constant rate pump with an accumulator and relief valve, Figure 4.5. The efficiency of this type of hydraulic power supply system showed some improvement because some energy is stored in the accumulator.

Figure 4.6 shows another type of hydraulic power supply system that employs a large-capacity accumulator, constant rate pump, and safety valve. The safety valve sets the highest safe pressure to protect the system. When the pump pressure is higher than the setting maximum pressure value, the

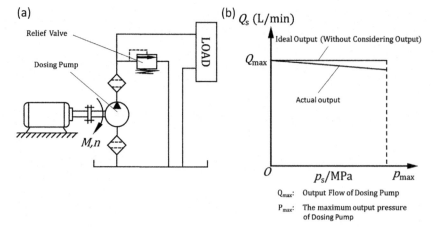

FIGURE 4.4   Early constant rate pump plus relief valve form. (a) Schematic diagram of hydraulic power system, (b) characteristic curve of hydraulic power system.

two-position, two-way electromagnetic valve engages to make the safety valve unload the flow. At that moment, the accumulator maintains the system pressure. When the system pressure is lower than the setting pressure, the two-position two-way electromagnetic valve operates to supply the power to the system, whereas the accumulator is charged at the same time. In this type of hydraulic power supply system, the main power loss is due to the leakage.

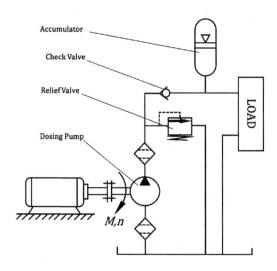

FIGURE 4.5   Constant rate pump plus relief valve plus accumulator.

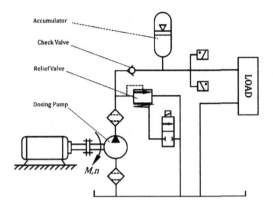

FIGURE 4.6   Schematics based on a large-capacity accumulator plus constant rate pump plus safe valve.

Some of the disadvantages of this type of the hydraulic power supply system include the need for a large-capacity accumulator and bigger pressure ripple, as shown in Figure 4.7. Thus, this hydraulic power supply system is not suitable in high-performance applications.

In the 1960s, the constant pressure variable pump sources, Figure 4.8, began to find practical applications. The main component of this pump system is a constant pressure variable pump and its regulation system. This type of hydraulic system can adjust the outlet pressure according to the load, resulting in improved efficiency. The principle of this hydraulic system is to control the angle of the swashplate by comparing the outlet pressure with spring setting pressure. When the pump outlet pressure is lower than the spring setting pressure, the angle of the swashplate increases and the outlet pressure increases accordingly. On the other hand, a decrease in the swashplate angle will result in the outlet pressure. Because the output flow is related to the system requirements, the constant pressure variable pump system has high efficiency.

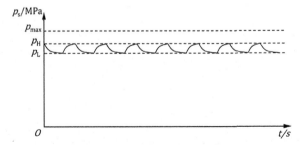

FIGURE 4.7   The pressure ripple of a large-capacity accumulator plus constant rate pump plus safe valve form.

(a)

(b)

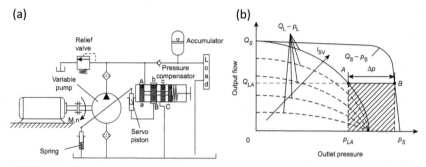

FIGURE 4.8 Constant pressure variable pump source schematics. (a) Structure diagram, (b) flow rate versus pressure.

Figure 4.8(b) shows the relation between the flow rate (Q_L) and pressure (p_L) of a constant pressure variable pump. Here, i_{sv} is the current in the servo valve, whereas direction of the arrow indicates increasing current. The solid line expresses the load flow characteristic curve under the rated current. When the load operates at point A and the pump provides the fluid at point B, the shaded area in the diagram is the power loss. The power loss will generate heat and influence the system stability.

The load-sensing pump, also called power-matching pump, Figure 4.9, appeared in the 1970s. Unlike the constant pressure variable pump, the

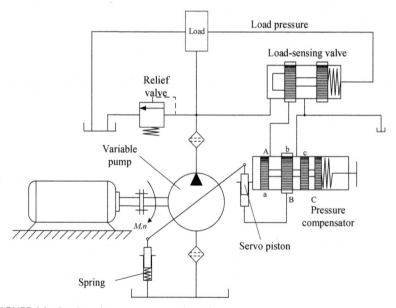

FIGURE 4.9 Load-sensing pump source schematics.

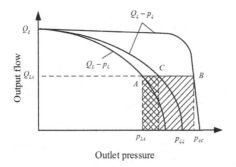

FIGURE 4.10   Comparison between a load-sensing pump and a constant-pressure variable pump.

load-sensing pump can maintain the constant pressure difference between pump outlet pressure and load pressure through load-sensitive valve feedback.

Figure 4.10 shows the comparison diagram between a load-sensing pump and a constant pressure variable pump. When the load operates at point A, the constant pressure variable pump operates at point B and the load sensing pump operates at point C. The shaded area is the power loss of constant pressure variable pump and the latticed area is the power loss of the load-sensing pump. It is apparent that the efficiency of the load-sensing pump is much better than the constant-pressure variable pump.

The advantage of a load-sensing pump is that the pump can better match the load requirement. The disadvantage is that it needs an extensive load-sensing circuit that can easily lead to a positive feedback phenomenon. In addition, the output of a load-sensing pump depends on the pressure difference between the load pressure and the load-sensitive valve, resulting in a tradeoff between performance and energy efficiency.

4.3 INTELLIGENT HYDRAULIC POWER SUPPLY SYSTEM

To reduce power losses in hydraulic systems, Rockwell Corporation conducted statistical analyses of flight profiles for various fighter aircraft's hydraulic systems [14]. The results indicated that the time during which an aircraft operates in the highest pressure regime (8000 psi (56 MPa)) is less than 10% of the total flight time, and that pressure of 3000 psi (21 MPa) is sufficient for the remainder of the flight time. Table 4.2 shows the statistical result of a military aircraft flight regimes. [21,22].

A dual-pressure pump, a combination of double-pressure variable pumps, was presented in the United States and the United Kingdom. During flight, a dual-pressure pump operates in a high-pressure regime when required; otherwise, the dual-pressure pump operates in a low-pressure regime. The United States adopted the 3000 psi/5000 psi (21 MPa/35 MPa) dual-pressure

TABLE 4.2 Statistical List of Flight Regimes [21,22]

No.	Flight Condition	Flight Time/min	Percentage of Total Flight Time/%	Flight Altitude/km	Mach Number/Ma
1	Take off	3	1.9	S.L.	0.28
2	Climb and cruise	48	29.6	35 K	0.8
3	Search and fall	36	22.2	30 K	0.7
4	Assault	4	2.4	S.L.	1.1
5	Fighting	5	3.2	10 K	0.6
6	Cruise and descent	48	29.6	40 K	0.8
7	Landing	18	11.1	S.L.	0.28
Total		162	100		

S.L. - sea level.

pump in the F/A-18E/F [17]. Figure 4.11 shows the flow rate—pressure curve of a dual-pressure pump.

When the load operates at point A_1, the dual-pressure pump provides the fluid at point B_1. When the load operates at point A_2, the dual-pressure pump provides the fluid at point B_2. The shaded area indicates power loss. Compared with the constant pressure variable pump, the efficiency of the dual-pressure pump shows great improvement because it is better matched with the external load. However, there is still some power loss in the system. If the

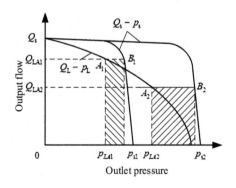

FIGURE 4.11 Flow-pressure curve of the dual-pressure pump source.

system operates between low- and high-pressure, the power loss is still considerable.

In the 1980s, engineers started to conduct research on intelligent variable pressure pumps, which adopt various sensors and microprocessor to control the pump outlet pressure and flow rate according to the system requirements. This type of pump can calculate the required pressure and flow rate according to the flight profile in an embedded microprocessor and control the operational mode of the pump to provide the appropriate pressure-flow rate output. Because the intelligent hydraulic power supply system can provide fluid according to the flight profile requirement, the power loss can be decreased and the weight/volume can be reduced. Abex Corporation did some experiments on the F-15 "Iron Bird" [15]. The results indicated that the intelligent pump decreases the consumed power by 39% and the discharge fluid temperature decreases between 18 and 24 °C, so the intelligent hydraulic power supply system has a bright prospects for future application [21]. Figure 4.12 shows the intelligent hydraulic power supply system and its interface with flight control system as developed by Vikers Corporation and McDonnell Douglas Corporation [22].

Figure 4.13 compares pump temperature, heating power, and consumed power between an intelligent pump and a constant pressure variable pump in the ground experiment. The results indicate that the heating power is significantly reduced in the intelligent pump [23].

4.3.1 The Requirement of an Intelligent Control Pump

General Aircraft specifications of a hydraulic pump requires that the pump be tested for maximum instantaneous pressure, response time, and settling time. The response time should be less than 0.02 s if the switching valve changes the pump output flow. When the pump changes the outlet pressure from full flow pressure to zero flow pressure, or vice versa, the transient peak pressure should not exceed 135% of the rated pressure, the response time should not be longer than 0.05 s, and the settling time should be less than 1 s (as shown in Figure 4.14). Although there are no special requirements for an intelligent control pump, aircraft hydraulic systems will have higher dynamic requirements for an intelligent pump.

There are two types of hydraulic systems in aircraft: a hydraulic transmission system and a hydraulic servo system. One of the most important tasks for the hydraulic transmission system is to maintain high efficiency. Hence, the intelligent pump should deliver variable flow function according to the required load.

The ideal pump for a hydraulic servo system is one with the constant pressure pump; however, its efficiency is relatively low. Meanwhile, in the typical aircraft flight profile, the flight time requiring high pressure is less than 10% of the total time of flight, the rest of the time it only operates at low pressure.

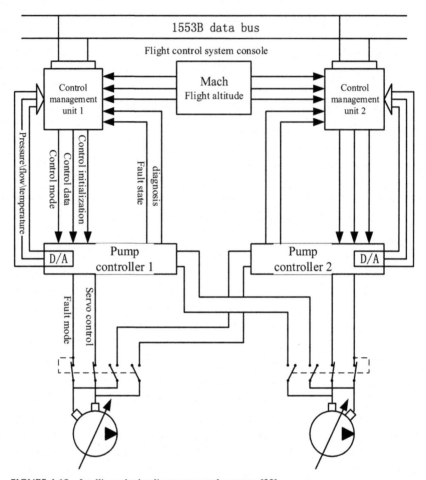

FIGURE 4.12   Intelligent hydraulic power supply system [22].

Therefore, intelligent control pumps should have a variable pressure capability, as shown in Figure 4.15.

Since the power that the engine provides to hydraulic system is constant, the hydraulic power could not exceed the maximum value. To provide the hydraulic power in different flight states (altitude, speed), its power should have certain limitation. Therefore, intelligent control pumps should have a variable power function that is within safety limits. Once an aircraft's hydraulic system power reaches a certain value (changes with different state of the aircraft flight), intelligent control reverts back to a constant power mode, as shown in Figure 4.16.

The pressure drop of a control valve should be considered when specifying a hydraulic servo system. The intelligent pump should reduce the throttling pressure drop as much as possible to improve efficiency. A load-sensitive

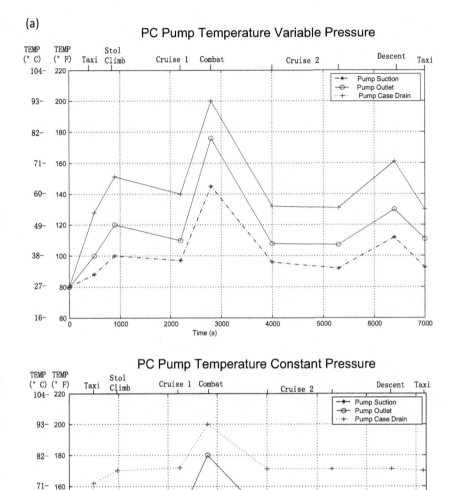

FIGURE 4.13 The comparison diagram between an intelligent variable pump and a constant pressure variable pump. (a) Temperature comparison diagram, (b) heating power comparison, and (c) consumed power comparison.

(b)

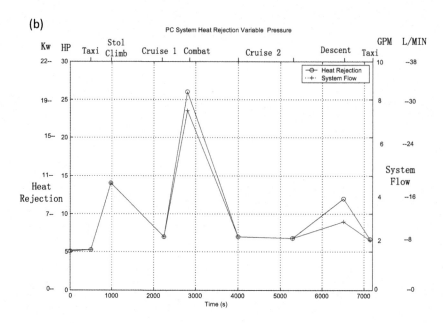

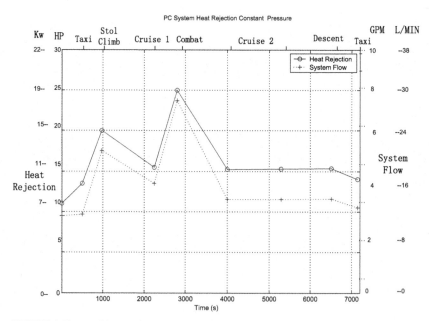

FIGURE 4.13 cont'd

(c)

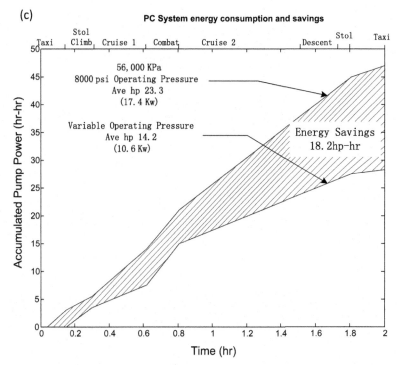

FIGURE 4.13   cont'd

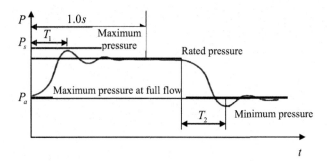

FIGURE 4.14   General constant pressure variable pump dynamic requirement.

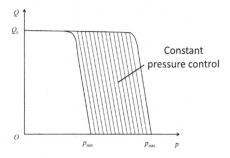

FIGURE 4.15   Variable pressure regulation characteristics.

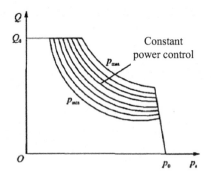

FIGURE 4.16 Constant power regulation characteristics.

pump can satisfy this requirement, so an intelligent control pump should have the load-sensitive mode, as shown in Figure 4.17.

Since different flight states require different pressures or flow rates from a pump, the performance of an intelligent control pump can be obtained according to the flight states. For example, Figure 4.18 shows the relation among flight altitude, Mach number, and the corresponding pressure. Through detecting the altitude and Mach number of aircraft, the corresponding operational pressure can be obtained as a pressure control signal.

4.3.2 The Structure of an Intelligent Variable Pump

Generally, aircraft require the following functions of intelligent hydraulic power supply system:

1. Report the correct states of hydraulic power supply system
2. Determine the operational modes automatically and according to the aircraft flight state and external load
3. Satisfy the requirement of the flight actuation system in the full flight profile including maneuverability and stability
4. Have a redundant design to keep aircraft operation reliable.

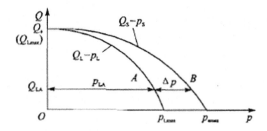

FIGURE 4.17 Load-sensitive adjustment feature.

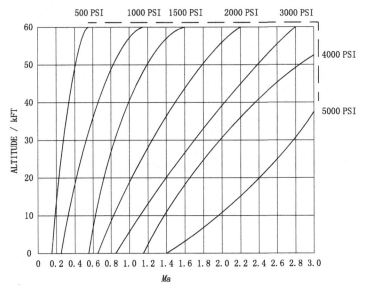

FIGURE 4.18   The relation among flight altitude, Mach number, and the pump pressure.

To meet these requirements, the intelligent variable pump is designed as shown in Figure 4.19, where the pressure-compensated valve is used to provide feedback to the outlet pressure of pump. When the outlet pressure of pump is greater than the spring setting pressure, the feedback force drives the cylinder to adjust the swashplate angle to realize the variable.

The system shown in Figure 4.19 has a servo valve parallel to the pressure-compensated valve, which is composed of a control system to control the swashplate angle with auxiliary pump and cylinder. The servo valve is controlled by the onboard controller of the intelligent variable pump, and the states of pump are detected by the installed sensors. The pressure sensor is installed in the output port of pump, whereas the return oil pressure sensor is installed in the oil-return port. The linear variable differential transformer is used to measure the displacement of cylinder, obtain the angle of swashplate, and then calculate the output flow rate. In addition, there are temperature sensors installed in the output port, shell, and return port. The controller of intelligent hydraulic power supply system can integrate the flight states parameters, flight mission requirement, and pump states and determine the operational modes and control signal. In the pressure-compensated valve circuit, there is a normally closed solenoid valve. If the servo valve fails, the controller of intelligent pump will open the solenoid valve and the intelligent variable pump becomes a constant pressure variable pump.

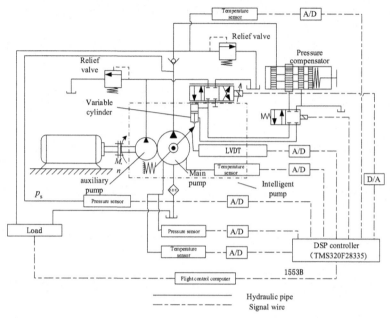

FIGURE 4.19   Intelligent variable pump structure [69].

4.3.3 Information Interaction Analysis of an Intelligent Control Pump

Since the flight control system consists of flight computer, hydraulic power supply, and actuator, it is necessary to develop information interaction between the flight control computer and controller of intelligent control pump.

1. The flight control computer and actuator are composed of a closed-loop flight control system in which the flight computer gives the control signal to drive the actuator movement. The displacement of the actuator is fed back to the flight control computer.

2. The flight control computer does not control the intelligent control pump but rather provides the system requirement to the pump. The controller of an intelligent control pump determines the control command and controls the intelligent pump autonomously and according to the flight requirement. This requires fast transmission of a large amount of data between the flight control computer and controller of intelligent pump (e.g., flight speed, flight altitude, actuator states). The flight control computer is responsible for monitoring the states of the intelligent variable pump and for switching it to the constant pressure variable pump mode when the intelligent control pump fails.

3. The controller of the intelligent pump and main pump consists of a closed-loop system in which the controller selects the operational modes according to the flight profile and determines the control signal to control the servo valve. The controller of intelligent pump needs to collect six channel data including three temperature channels, two channel pressures, and one channel displacement. The intelligent pump controller provides one control signal to the servo valve.

The information transmission between flight control computers and the controller of the intelligent control pump is shown Figure 4.20.

In Figure 4.20, various parameters of an intelligent pump can be sent to the flight control system by its controller. The system can diagnose the fault for each pump, and if the pump were to fail, the system can isolate it and switch on the backup systems to avoid a major accident.

The detailed operation process is as follows:

1. The flight control computer generates the operational cycle of mission profile for intelligent variable pump according to the flight states and control surfaces load.
2. In each operational cycle, the flight control computer transmits the command to the actuator and the intelligent pump controller.
3. In the operational cycle, the intelligent pump controller determines the operational modes and controls the servo valve according to the flight control computer command.
4. The servo valve controls the cylinder motion to change the swashplate angle, whereas the intelligent control pump delivers the pressure and flow rate required for the actuation system.

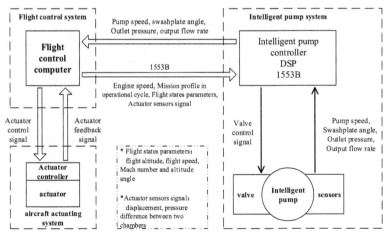

FIGURE 4.20 Information interaction between a flight control computer and controller of the intelligent pump.

TABLE 4.3 Related Information Between Two Systems

Transmitting Terminal	Receiving Terminal	Transmission Information
Flight control computer	Intelligent pump controller	Engine speed, mission profile in operational cycle, flight states parameters (flight altitude, flight speed, Mach number, and altitude angle)
Intelligent pump controller	Flight control computer	Sensors data of intelligent pump (pump speed, swashplate angle, outlet pressure, and output flow rate)
Flight control computer	Actuation system	Control command of actuator
Actuation system	Flight control computer	Actuator sensors signal (displacement, pressure difference between two chambers)
Actuation system	Intelligent pump controller	Actuator sensor signal (displacement, pressure difference between two chambers)

5. The intelligent control pump consists of four operational modes including pressure modes, flow rate mode, power mode, and load-sensitive mode.
6. The next operational cycle starts after the completion of the current operational cycle.

The related information between the flight control computer and intelligent pump controller is listed in Table 4.3.

4.3.4 Control of Intelligent Pump

4.3.4.1 The Control Modes of Intelligent Control Pump

The controlled command determined from the flight profile shown in Figure 4.21. The required hydraulic pressure and flow rate can be calculated according to the specified flight profile.

In the hydraulic power supply condition, a hydraulic system provides the required flow rate to actuation system and other users, so the intelligent control pump needs the flow rate mode to provide high-pressure fluid to the load.

The hydraulic servo control condition is the best way to adopt the constant pressure power supply system. The constant pressure power supply is designed according to the maximum operational condition, whereas the aircraft operates in this condition for a very short time. It is necessary for the intelligent control pump to be able adjust its pressure according to the requirement, viz. pressure mode.

The maximum power that the engine provides to a hydraulic system is constant when the aircraft flies at a certain altitude and a certain speed.

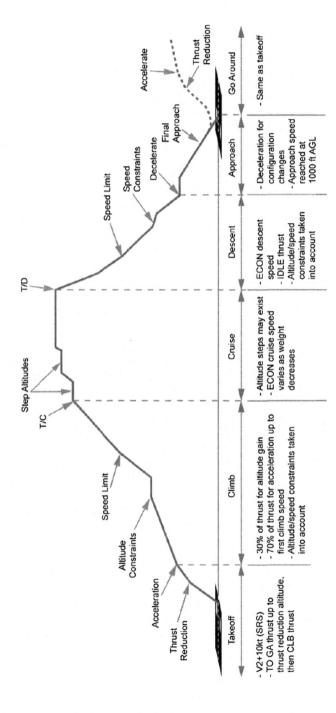

FIGURE 4.21 Typical flight profile of a commercial aircraft.

TABLE 4.4 Four Modes of Intelligent Control Pump

Operational Mode	Switch Condition	Control Mode	Applications
Variable flow rate mode	Small load, big flow rate, and need constant speed manipulation	Constant flow rate control	Retraction landing gear, retraction flap, retraction speed brake
Variable pressure mode	Common	Constant pressure control	Full flight profile
Variable power mode	Required power over the limit	Constant power control	High maneuverability over the limit
Load-sensitive mode	Load change frequently	Load-sensitive control	Energy-saving control under normal conditions

Aircraft safety could be affected if the power required from the engine exceeds the maximum supply, so the flight control system sets a threshold according to the power of engine. If the required hydraulic power exceeds the threshold during the flight, the intelligent controls pump switches to the power mode. Under these circumstances, the intelligent control pump operates in the power mode and provides hydraulic power that does not exceed the threshold.

In the hydraulic servo control condition, an intelligent control pump can operate in a load-sensitive mode by controlling the pressure drop of the servo valve to improve the efficiency of hydraulic system.

In summary, the intelligent control pump has four operational modes suitable for different loads and energy-saving requirements, including pressure mode, flow rate mode, power mode, and load-sensitive mode, summarized in Table 4.4.

4.3.4.2 Mode Selection and Switching of an Intelligent Control Pump

During the flight process, the pump states (speed, inlet temperature, outlet flow rate, leakage flow rate, outlet pressure) of the pump are detected by sensors and are transmitted to the intelligent pump controller. The flight control computer determines the operational mode for the pump controller according to the flight states (flight altitude, speed, etc.), pump states, and load states

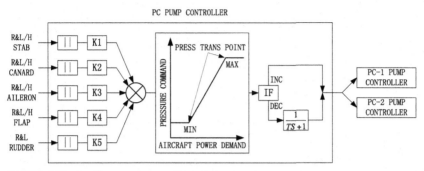

FIGURE 4.22 Intelligent pump controller switch.

(pressure, speed, displacement, and angle). The intelligent pump controller controls the operational mode by flight control computer command. If the hydraulic pump fails, the pump controller can isolate the fault pump and switch to another pump as shown in Figure 4.22.

Figure 4.22 shows the principle of intelligent pump controller, where two pumps provide the high-pressure fluid to the load. The pump controller receives the command from the flight control computer and determines the target pressure value for the hydraulic pump. If the target pressure is greater than the current operating pressure, the command control signal is sent to the pump to increase pressure; if the target pressure is lower than the current operating pressure, the command control signal is sent to the pump to decrease pressure.

4.3.4.3 Mathematical Model of Intelligent Pump

The flow rate in the variable flow rate mode of intelligent pump can be described as

$$Q_t = 1.61 \times 10^{-4} n\gamma = \frac{1.61 \times 10^{-4} n\left[\frac{K_q}{A_c L}x_v - \frac{K_e}{A_c^2 L^2}\left(\frac{V_c}{4\beta_e K_{ce}}S + 1\right)V_c p_s\right]}{S\left(\frac{S^2}{\omega_h^2} + \frac{2\xi_h}{\omega_h}S + 1\right)} \tag{4.4}$$

where Q_t is the flow rate of pump (m³/s); $q_t(\gamma) \approx 1.61 \times 10^{-4}$ is the angular displacement (m³/(s·rad)); γ is the angle of swashplate (rad); x_v is the spool displacement of servo valve in swashplate variable mechanism (m); K_q is the flow amplification coefficient (m²/s); A_c is the piston area of cylinder (m²); L is the distance between variable cylinder axis to the center of swashplate (m); p_s is the outlet pressure of pump (Pa); K_e is the total pressure-flow coefficient (m⁵/(N·s)); V_c is the total volume of variable cylinder (m³); β_e is the fluid equivalent modulus of elasticity (N/m²); V_p is the coefficient of pressure disturbance torque (m^3); and ω_h is the inherent frequency of swashplate variable mechanism (rad/s):

$$\omega_h = \sqrt{\frac{4\beta_e A_c^2 L^2}{V_c J_c}} \qquad (4.5)$$

and ξ_h is the relative damping coefficient of swashplate variable mechanism:

$$\xi_h = \frac{K_e}{A_c L} \sqrt{\frac{\beta_e J_c}{V_c}} \qquad (4.6)$$

where J_c is the moment of inertia of swashplate variable mechanism ($kg \cdot m^2$).

The mathematical model of intelligent pump at variable pressure mode can be expressed as

$$p_s = \frac{1}{C_{pl}\left(1 + \frac{S}{\omega_S}\right)} \left[\frac{1.61 \times 10^{-4} n \left[\frac{K_q}{A_c L} x_v - \frac{K_e}{A_c^2 L^2}\left(\frac{V_c}{4\beta_e K_e} S + 1\right) V_p p_s\right]}{S\left(\frac{S^2}{\omega_h^2} + \frac{2\xi_h}{\omega_h} S + 1\right)} - |Q_L| \right]$$

$$(4.7)$$

where C_{pl} is the total leakage coefficient ($m^5/(N \cdot s)$) and ω_S is the volume lag tuning frequency (rad/s):

$$\omega_S = \frac{\beta_e C_{pl}}{V_S} \qquad (4.8)$$

where V_S is the volume of pump outlet (m^3) and $|Q_L|$ is the flow that load receives from the pump (m^3/s).

The mathematical model of intelligent pump at the variable power mode can be represented as

$$W_0 = \frac{1}{C_{pl}\left(1 + \frac{S}{\omega_S}\right)} \left[\frac{1.61 \times 10^{-4} n \left[\frac{K_q}{A_c L} x_v - \frac{K_e}{A_c^2 L^2}\left(\frac{V_c}{4\beta_e K_e} S + 1\right) V_p p_s\right]}{S\left(\frac{S^2}{\omega_h^2} + \frac{2\xi_h}{\omega_h} S + 1\right)} - |Q_L| \right]$$

$$\times \frac{1.61 \times 10^{-4} n \left[\frac{K_q}{A_c L} x_v - \frac{K_e}{A_c^2 L^2}\left(\frac{V_c}{4\beta_e K_e} S + 1\right) V_p p_s\right]}{S\left(\frac{S^2}{\omega_h^2} + \frac{2\xi_h}{\omega_h} S + 1\right)}$$

$$(4.9)$$

The mathematical model of pump in the load-sensitive mode is similar to Eqn (4.7), in which the real output pressure, load pressure p_L, and load flow rate Q_L should be fed back.

According to this relationship, the block diagram of intelligent control pump in different kinds of operational modes can be shown in Figure 4.23 [24].

In Figure 4.23, Δp_u (Pa) is the pressure drop of servo valve and $G_v(S)$ is the transfer function of servo system controller.

Applying AMEsim software, some intelligent pump simulation results are shown in Figures 4.24–4.26.

The simulation results indicate that the intelligent pump can track the command signal well in the variable flow rate mode and under different load requirements.

Figure 4.25 indicates that the intelligent pump can adjust the pressure from 3000 psi (21 MPa) to 4000 psi (28 MPa) continuously. It can be seen that there are some oscillations in the high-pressure area.

The simulation considers two switching events: from the pressure mode to the flow rate mode and then from the flow rate mode to the pressure mode. It is apparent that the intelligent pump can switch from the pressure mode to the flow rate mode smoothly.

4.4 NEW ARCHITECTURE BASED ON 2H/2E

In 1936, Douglas DC3 initially used one engine-driven pump and two hydraulic circuits to provide the hydraulic power to booster and other users. At that time, the pilot manipulated the mechanical transmission device to control the surfaces, and aircraft flight control systems adopted the mechanically signaled hydraulic powered actuator. In 1953, aircraft started to adopt two pumps driven by two engines, and the electrically signaled hydraulic powered actuator appeared with electro-hydraulic servo valve to realize the stability augmentation, control augmentation and fly-by-wire [25]. For nearly half a century, aircraft control surfaces were hydraulically actuated [26]. To achieve high reliability, current aircraft have adopted three to four redundant hydraulic networks to supply power to electrically signaled hydraulically powered actuator (HA). Because this type of actuator depends on the centralized hydraulic power supply system, its installation and maintenance are expensive, and its weight results in increased fuel consumption [27]. Engineers found a way to replace hydraulic power transmission by electrical power transmission (power-by-wire) [28]. This type of design can reduce the overall system weight, increase power efficiency, improve the reliability, and have fail-safe modes. In 2006, for the first time, A380 introduced two electrical power supply systems (2E) and two hydraulic power supply systems (2H) to aircraft flight controls (shown in Figure 4.27). Then the electrically signaled electrically/hydraulically powered actuator appeared such as electro-hydrostatic actuators (EHA) and electro-backup hydraulic actuator (EBHA) shown in Figure 4.28. The electric actuation offers some advantages over centralized hydraulics such as fewer leaks, easy removal of components without

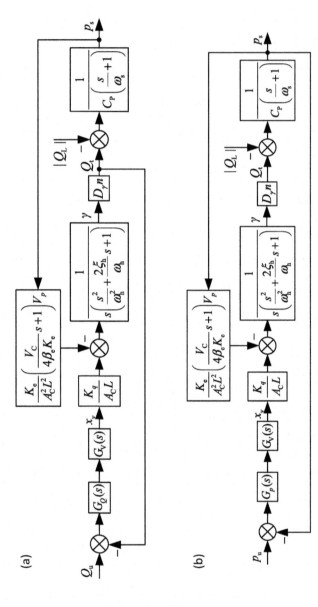

FIGURE 4.23 Block diagram of intelligent control pump modes. (a) Flow rate control mode, (b) pressure control mode, (c) power control mode, and (d) load-sensitive control mode.

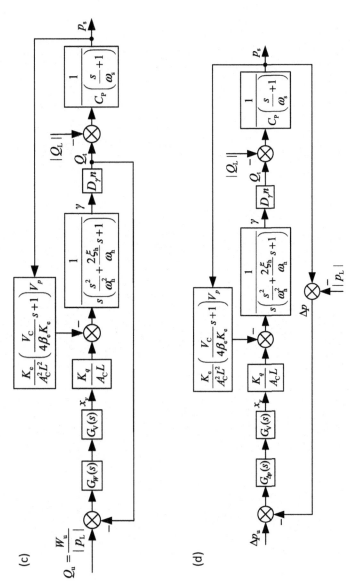

FIGURE 4.23 cont'd

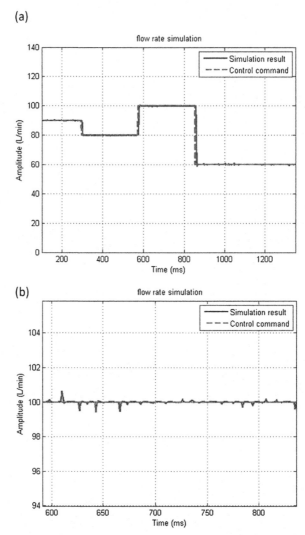

FIGURE 4.24   Intelligent pump simulation results in variable flow rate mode. (a) Flow rate control mode, (b) flow pulsation.

breaking into hydraulic lines, and easier physical separation of redundant electrical systems compared with hydraulic systems (Figures 4.27 and 4.28).

The corresponding actuators consist of conventional hydraulic servo control actuator (HA), EHA, and EBHA.

4.4.1 EHA Principle

Since the late 1970s, engineers have focused on development of the power-by-wire system with the goal of reducing the weight of aircraft control system [25].

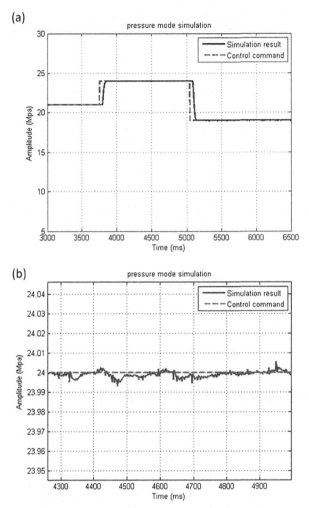

FIGURE 4.25   Intelligent pump simulation result in variable pressure modes. (a) Pressure control mode, (b) pressure pulsation.

It was discovered that a distributed hydraulic actuator can effectively minimize required pipes between the centralized hydraulic power supply system and the actuator while improving installation and maintenance. EHA consists of an electric motor, pump, and actuator ram requiring about 1 pint of hydraulic fluid. EHA systems are power-by-wire actuation systems that use aircraft electric power and self-contained hydraulic system for flight control surface actuation. The evolution toward more electric aircraft replaces some hydraulic power networks with an electric network. The A380 actuation system adopts two hydraulic power supply systems and two electric power supply systems. The main component of increasingly electrical actuation system is based on the

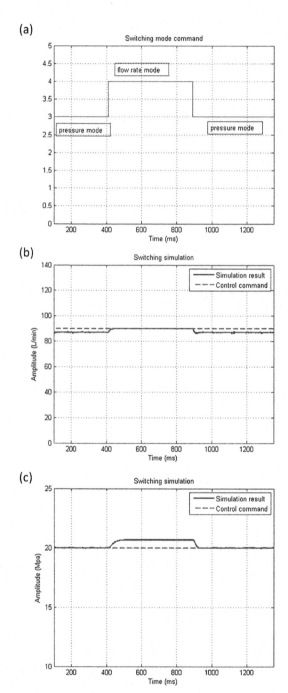

FIGURE 4.26 Mode switching simulation between the pressure mode and the flow rate mode. (a) Mode switching command, (b) flow rate tracking under flow switching, and (c) pressure tracking under pressure switching [24].

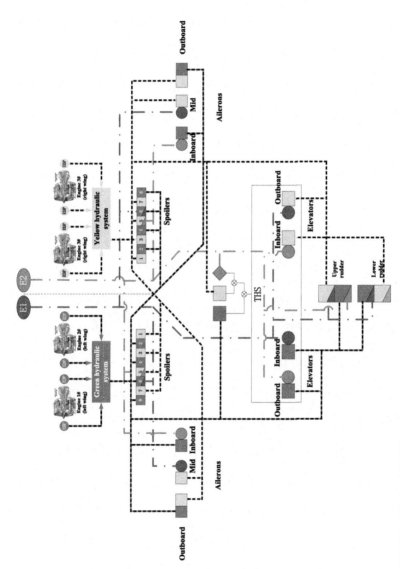

FIGURE 4.27 2H+2E architecture of A380.

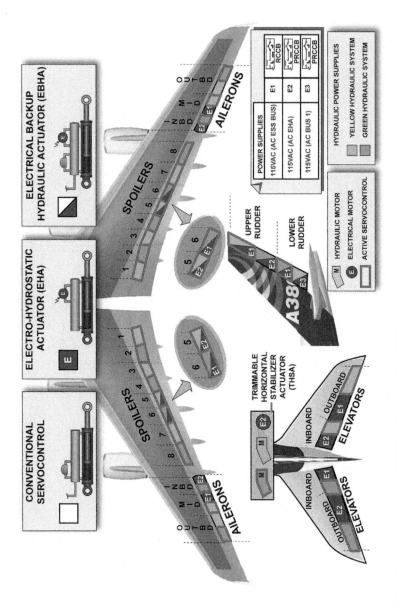

FIGURE 4.28   The new actuator in the 2H+2E architecture of the A380.

electrical motor driven pump directly connected to hydrocylinder. Since there are no external hydraulic connections, the overall weight is significantly reduced.

There are three types of EHA according to the operational conditions of a motor and pump; each is described as follows:

- EHA-FPVM: where FPVM means fixed pump displacement and variable motor speed. In this type of EHA, Figure 4.29, the swashplate angle is fixed and the flow rate is controlled by adjusting the motor speed.

The check valve and accumulator, Figure 4.29, consist of the closed type reservoir, which compensates for the fluid loss due to leakage and keeps the lowest pressure to provide cavitation. The function of a relief valve is to provide protection when the system pressure exceeds the pressure limit. A bypass valve is used to isolate the fault. In order to meet the requirements of driving load, the power of electrical motor should be designed a certain margin to cover the load power completely since pump discharge is constant.

- EHA-VPFM: where VPFM means variable pump displacement and fixed motor speed. In this type of EHA, Figure 4.30, the speed of motor is fixed and the flow rate is controlled by the servo valve and cylinder through adjustment of the angle of the swashplate of pump.

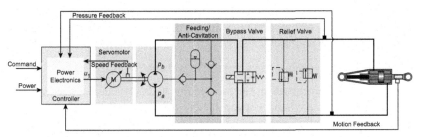

FIGURE 4.29   Fixed pump displacement and variable motor speed EHA [25].

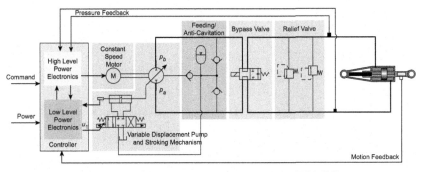

FIGURE 4.30   Variable pump displacement and fixed motor speed EHA [25].

This type of EHA provides high speed for the motor even under no-load or small-load conditions, so it will result in energy waste. In addition, its variable mechanism is complex and heavy.

- EHA-VPVM: in this type of EHA, Figure 4.31, the speed of motor and the displacement of pump can be controlled at the same time.

This type of EHA not only can improve the system's dynamic performance, but can also keep the motor power matching the system requirement to improve the system rigidity. This two variable control mode is expected to achieve optimal performance but the structure complexity increases while the reliability decreases.

Since the more electrical aircraft requires the actuator to be simple and light, all modern aircraft have adopted EHA-FPVM. The key technologies of EHA include

- High-performance brushless DC motor and drive technology

The flow and pressure of EHA are controlled by a motor-pump module. It requires the high-efficiency, high-power-density brushless DC motor.

- Integrated motor-pump-cylinder system

Displacement of a traditional hydraulic pump is relatively big and the pump operates under a one-directional constant speed working condition, all of which result in long operational life. EHA requires the pump to operate in a bidirectional mode, with a wide range of speed, which has significant influence on its performance and useful life.

- Thermal management and energy-saving design

Since the EHA integrates an electrical motor, hydraulic pump, cylinder, tank, and accumulator in a compact space, its thermal problem is prominent. Thus, decreasing the thermal generation of motor and pump and optimization of thermal radiation capability of accumulator is a very important issue for EHA.

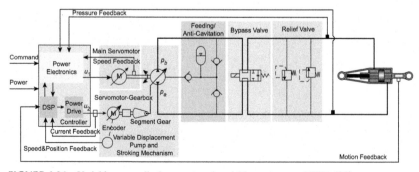

FIGURE 4.31   Variable pump displacement and variable motor speed EHA [25].

- Control strategy of EHA

EHA consists of an electrical motor, pump, and cylinder, Figure 4.32. Increasing the number of components in the systems increases the order of the system and results in complex behavior. The frequency and position stiffness of electrical motor are low, which makes it susceptible to external disturbance. Nonlinear control should be considered for the EHA control strategy.

4.4.1.1 Model of Brushless DC Motor

The electrical motor in EHA adopts the brushless DC motor, whose electromotive force balance equation can be described as [29]:

$$u_e = K_e\omega_e + L_e\frac{di_e}{dt} + R_e i_e \tag{4.10}$$

where u_e is motor control voltage, i_e is motor current, K_e is opposing electromotive force coefficient, ω_e is motor speed, R_e is armature resistance, and L_e is armature inductance.

The torque balance equation of motor can be expresses as

$$K_m i_e = T_e + J_m\frac{d\omega_e}{dt} + B_m\omega_e \tag{4.11}$$

where K_m is electromagnetic torque constant, T_e is output torque of motor, $B_m = B_e + B_p$ is the total load damping coefficient of motor-pump module, and $J_m = J_e + J_p$ is the total moment of inertia of motor-pump module.

4.4.1.2 Model of a Hydraulic Pump

Assuming that the motor is rigidly connected with the pump and that the efficiency of pump is 100%, then the input torque of pump can be described as

$$T_e = \frac{Vn(P_a - P_b)}{\omega} = \frac{V(P_a - P_b)}{2\pi} = V_P(P_a - P_b) \tag{4.12}$$

where V is the displacement of pump, P_a and P_b are the outlet pressure and inlet pressure, respectively, $V_P = V/(2\pi)$. Assuming that the leakage of pump is zero, then the flow rate of pump is:

$$q_a = q_b = Vn = V_P\omega_e \tag{4.13}$$

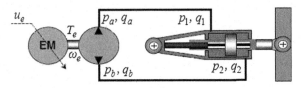

FIGURE 4.32　Diagram of a typical EHA.

4.4.1.3 Model of Cylinder

In Figure 4.32, the inlet pressure and flow rate of cylinder are P_1 and q_1, respectively, the outlet pressure and flow rate of cylinder are P_2 and q_2, respectively. Without considering the pressure drop between pump and cylinder, the load pressure P_e and flow rate Q_e are

$$\begin{cases} P_e = P_1 - P_2 = P_a - P_b \\ Q_e = (q_1 + q_2)/2 = (q_a + q_b)/2 \end{cases} \tag{4.14}$$

Assuming that the fluid temperature is constant and that the leakage of cylinder is laminar, the continuous flow equation of cylinder can be described as [30]

$$V_p \omega_e = A_e \frac{dx_e}{dt} + \frac{V_e}{4E_e} \frac{dP_e}{dt} + C_{el} P_e \tag{4.15}$$

where A_e is the effective area of piston, x_e is the displacement of piston, V_e is the total volume of cylinder, C_{el} is the total leakage coefficient and $C_{el} = C_{eli} + 0.5C_{ele}$, C_{eli} and C_{ele} are inner and external leakage coefficient, and E_e is the bulk modulus of elasticity.

The force balance equation of cylinder is

$$A_e P_e = m_e \frac{d^2 x_e}{dt^2} + B_e \frac{dx_e}{dt} + F_e \tag{4.16}$$

The transfer function block diagram of EHA is shown in Figure 4.33.

Here the proportional-integral-derivative controllers are used in the current loop, velocity loop, and displacement loop. The step response and Bode plot are shown in Figures 4.34 and 4.35.

The rise time of EHA, Figure 4.34, is approximately 0.6 s and the overshoot of surface displacement is zero. This performance is the same as for HA. When the control surface is affected by the instantaneous disturbance, the EHA output has some oscillations. When disturbances disappear, the EHA output rapidly recovers and transitions to a new steady state.

The frequency width at -3 dB is 12 Hz, so the dynamic performance satisfies the requirement of an aircraft actuation system.

4.4.2 EBHA Principle

EBHA is another type of EHA, which integrates the HA and EHA in one module, Figure 4.36. Since the EBHA operates like EHA in a backup mode, it needs extra power electronics and a control box.

The EBHA has two operating modes:

- Normal mode—Hydraulic mode: In this mode, the EBHA receives power from the centralized hydraulic power supply system, and the servo valve controls the actuator to drive the surface movement according to the flight control computer command.

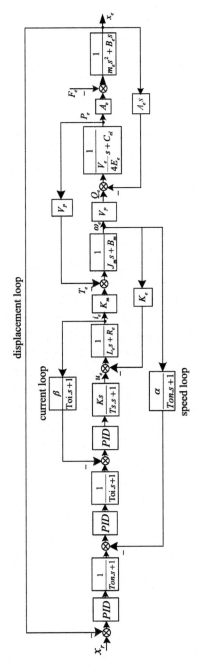

FIGURE 4.33 Block diagram of EHA.

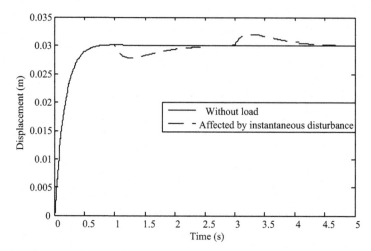

FIGURE 4.34   Step response of EHA.

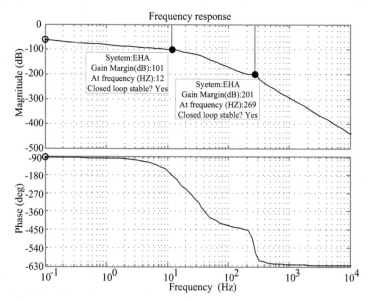

FIGURE 4.35   Bode diagram of EHA.

- Backup mode—EHA mode: In this mode, EBHA operates like EHA, which receives the power from the electrical power supply system. A flight control computer gives the command to the EHA control unit, and an electric motor drives the pump and cylinder of the control surface.

Since the EHA is integrated actuation system, it is highly energy efficient and provides an overall weight reduction benefit to the aircraft. The relationship among HAS, EHA, and EBHA is shown in Figure 4.37.

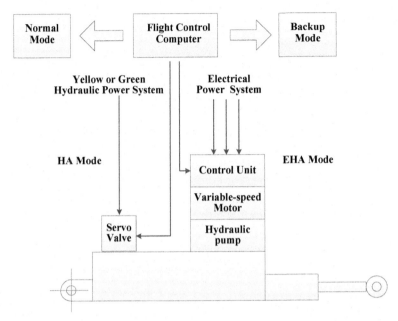

FIGURE 4.36   EBHA structure.

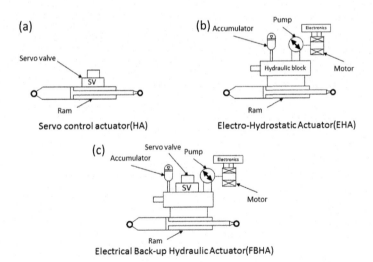

FIGURE 4.37   Relationship among HA, EHA, and EBHA.

4.4.3 Dissimilar Redundant Hybrid Actuation System

Current aircraft employ dissimilar redundant hybrid actuation systems based on HA and EHA, Figure 4.38, to improve the reliability of flight control system.

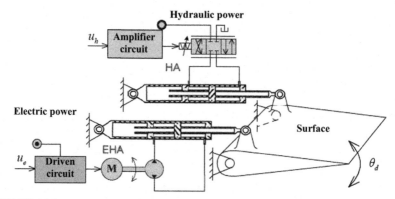

FIGURE 4.38   Dissimilar redundant hybrid actuation systems.

There are three modes of the hybrid actuation system:

- HA active/EHAS follow-up
- HA fault/EHA active
- HA active/EHA active

In the HA active/EHA active mode of the hybrid actuation system, the force fighting emerges because of the inconsistent motion between HA and EHA, which are required to simultaneously drive the same load. With the static and dynamic differential, the hybrid actuation system cannot achieve the same position even under the same position demand. The position difference leads to actuator's output force difference. Thus, these types of systems require development of the force equalization control strategies.

4.4.4 Electromechanical Principle

With the development of power electronics and permanent magnets, the electromechanical actuator (EMA) becomes very promising for aircraft flight control. The EMA is driven by an electric motor that converts electricity to mechanical force in order to drive the control surface movement, Figure 4.39.

The EMA employs a high-speed electric motor to produce the motor torque and rotary motion. The gear reducer transforms rotary motion of the

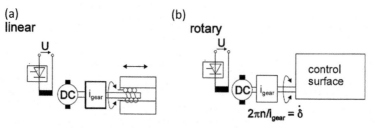

FIGURE 4.39   EMA structure. (a) Linear EMA, (b) rotary EMA.

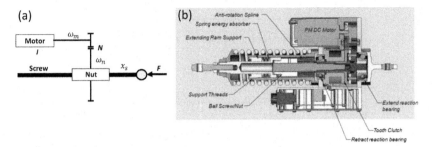

FIGURE 4.40   Principle of EMA. (a) EMA structure, (b) inside EMA.

motor to rotary motion of the screw. The rolling screw is used to convert the rotary motion to linear motion. The schematic diagram of EMA is shown in Figure 4.40 [30].

The torque of motor can be described as

$$T_m = I \frac{d\omega_m}{dt} \qquad (4.17)$$

where I is the rotor moment of inertia and ω_m is motor speed.

The gear reducer function is

$$\begin{cases} \omega_m = N\omega_n \\ T_m = \dfrac{1}{N} T_n \end{cases} \qquad (4.18)$$

where N is the gear reducer ratio.

The relation of crew/nut is

$$\begin{cases} \dfrac{dx_S}{dt} = \dfrac{p}{2\pi} \omega_n \\ T_n = \dfrac{p}{2\pi} F \end{cases} \qquad (4.19)$$

where p is crew pitch.

The force balance equation is

$$F = \left(\frac{2\pi}{p} N\right)^2 \frac{d^2 x_s}{dt^2} = m_e \frac{d^2 x_s}{dt^2} \qquad (4.20)$$

where m_e is the equivalent inertia.

An aircraft test indicates that the EMA can reduce flight control system weight by 25%, reduce maintenance by 42%, reduce mean time to repair (MTTR) by 50%, increase aircraft availability by 15%, and enhance the ballistic tolerance [30].

Although EMA has certain advantages, it also has some shortcomings such as jamming and seizing. As a result, EMA is used only in the secondary flight control for spoilers in the Boeing 787 and for engine thrust reversers in the A380.

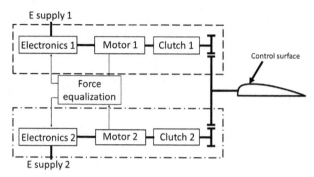

FIGURE 4.41   Torque summing redundant EMA.

To increase the reliability of EMA, there are two types of redundant actuation system designs based on the EMA.

1. Torque summing redundant EMA:
   Figure 4.41 shows the torque summing redundant EMA.
   This type of actuation system employs at least two actuators to drive the same load through a common gear. The redundant management of this actuation system should consider the following issues:
   a. Jamming
   b. Losing the transmission link
   c. Declutching the jammed path
   d. Avoiding force fighting through force equalization
2. Speed summing redundant EMA:
   Speed summing redundant EMA is shown in Figure 4.42, in which at least two actuators were employed to drive the same load through a speed summing device.
   The redundant management of this actuation system should consider the following issues:
   a. Jamming
   b. Losing the transmission link
   c. Detaching the jammed path
   d. Avoiding speed fighting through speed equalization

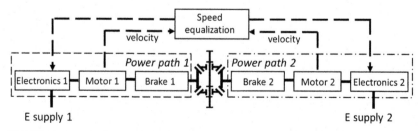

FIGURE 4.42   Speed summing redundant EMA.

4.5 PROGNOSTICS AND HEALTH MANAGEMENT OF HYDRAULIC SYSTEMS

Prognostics and health management (PHM) is an effective methodolgy to guarantee the high reliability, safety, and maintainability of aircraft through automatic detection, diagnosis, and prognosis of the adverse events during flight [31]. PHM represents a change in aircraft logistic support from the traditional fault detection based on sensors to intelligent prognosis, and from reactive response to active maintenance management based on the exact fault diagnosis. PHM changes corrective maintenance to condition maintenance and focuses not only on the fault diagnosis, but also emphasizes prognostics and repair strategy support. According to the Sandia National Laboratory, PHM has the capability to estimate the likelihood of a system failure over some future time interval so that appropriate actions can be taken.

A PHM system consists of the following elements:

- Raw data: Sensor data, historical maintenance/failure data
- Diagnostics: Data fusion, data interpretation
- Prognostics: Predictions of system "health"
- Health management: What should be done, when should it be done

PHM can improve aircraft safety through application of diagnostics and prognostics to fix faults before the critical event appears. It can also improve availability through better maintenance strategy and improve reliability through a more thorough understanding of the current health of the system and prognosis based maintenance. PHM can reduce total cost of maintenance through reduction of unnecessary maintenance and avoidance of unscheduled maintenance.

4.5.1 Development of PHM

Development of aircraft fault detection and diagnosis systems started in the 1950s [32]. The Palo Alto Corporation developed the first automatic testing and fault diagnosis device for aircraft engines [33,34]. In 1961, the National Aeronautics and Space Administration (NASA) started work on Project Apollo; during those early days of development, accidents would occur because of equipment failure. This prompted NASA in 1967 [35] to set up the mechanical failures prevention group to detect, diagnose, and predict failures. In the 1970s, the Built-In-Test (BIT) was adopted in the first generation of aircraft to conduct the fault diagnosis and isolation of aircraft electrical devices [36]. Afterward, Boeing 737s and A300s were equipped with BIT to perform fault diagnosis of analog electrical devices. The low fault diagnosis capability of BIT and its high alarm rate at the time presented great challenges to applying BIT systems. At the beginning of the 1980s, the Rome Air Development Center proposed the application of intelligent techniques to improve the efficiency and decrease the alarm rate of BIT systems. Boeing 757s and Airbus 310s [37] were equipped with this new system. At the end of the 1980s, a new fault diagnosis system with bus and distributed design was introduced and it was installed in Boeing 747s and Airbus 320s. This kind of system adopted integrated fault diagnosis and centralized fault display system to decrease system complexity and enhance the level of standardization [38]. In the 1990s, the requirements expanded from that of ensuring aircraft safety during flight to also providing the health condition, rapid diagnostics and location of the fault, prediction of the residual useful life, and providing a maintenance strategy. Then PHM was replaced by the intelligent reasoning and failure prediction system to improve the fault diagnosis capability, decrease the repair time, and improve the aircraft availability, operating benefit, and integrated supportability [38−40]. PHM was installed in Boeing 777s to provide the integrated capability of fault detection, allocation, isolation, prediction, and maintenance. NASA, Defense Advanced Research Projects Agency, and the Navy provided great contributions to applying PHM. At the end of the 1990s, the rapid development of computer networks, advanced microsensors, intelligent data fusion algorithms, and computer memory made it possible for real-time condition monitoring and integration of air-ground supportability [41,42]. At present, Honeywell, International, Impact, Lockheed Martin Aeronautics, and Moog Corporation provide the PHM system for aircrafts.

Significant contributors to PHM research, and development come from both academia and industry. Andrew Hess, Ian Barnard, Machael G. Pecht, and Ian K. Jennions are representative academic leaders in this area. Andrew Hess established the PHM society in 2009, and was recognized as the father of naval aviation propulsion diagnostics. Ian Barnard has been involved in the field of PHM for more than 30 years. He has consulted with more than 100 companies in the Pacific, North America, Asia, China, and Europe. These companies operate in heavy industrial sectors including defense, mining, power genera-tion, manufacturing, and process industries. Michael G. Pecht established the Center for Advanced Life Cycle Engineering at the University of Maryland and at the University of Hong Kong. Ian K. Jennions made contributions to the engine fault diagnosis and was a director of Integrated Vehicle Health Management center, which is funded by a number of industrial companies, including Boeing, BAE Systems, Rolls-Royce, Thales, Meggitt, MOD, and Alstom Transport. Company researchers Douglas W. Brown, Michael J. Roemer, Carl Byington of Impact Corporation, Philip A. Scandura of Honeywell Aerospace, Mark Scwabacher of NASA Ames, and Michael Candy and Keivin Line of Lockheed Martin Aeronautics are all considered to be industry leaders who made significant contributions the field of aircraft PHM methodologies.

PHM is a very promising methodology that can improve the reli-ability, safety, maintainability, and supportability of modern aircraft. PHM represents a change in aircraft logistic support. It changes the fault diagnosis from traditional BIT-based on sensors to intelligent fault prediction. It also changes the maintenance from traditional correction repair to condition-based maintenance. PHM not only focuses on the fault diagnosis but also emphasizes the failure prediction and maintenance strategy support. PHM can provide fault extension rule, predict the residual useful life according to the performance degradation and current state, and give the maintenance strategy.

PHM is currently used in F-35s, Boeing 787s, and Airbus 380s. Different aircraft have different PHMs. Helicopter uses the health and usage monitoring system and the space shuttle had adopted Integrated Vehicle Health Management. According to statistics from the Boeing Corporation, PHM enables airlines to save about 25% of the costs related to flight delay or cancellation and decreases the repair cost greatly.

4.5.2 PHM Structure

Aircraft PHM structure depends on system classification, restriction management among systems, system modes, and integrated strategy. PHM is an open system, which is portable, extensive, and cooperative, and interfaces with the information management system and logistic support management system [43]. PHM adopted the IEEE 1471-2000, ARINC604, ARINC624, and ISO 13374 to construct the PHM structure. Figure 4.43 shows the structure of PHM system, which has the following characteristics:

1. A hierarchical structure
2. Distributed and cross-platform structure
3. Open, modular, and standard interface specification
4. Can realize real time fault diagnosis, prognostics, and maintenance decision-making
5. Software facilitates safety

This kind of structure can provide the flexible environment with fault diagnosis and failure prediction, multiple reasoning algorithm integration, and real-time operation. It can also improve aircraft safety, increase reliability and availability, and reduce time to determine the health condition and maintenance costs.

An aircraft PHM system consists of three parts: airborne PHM, air-ground data link, and a ground maintenance system. Airborne PHM is in charge of monitoring health conditions in real time and carrying out the abnormal/diagnosis/prediction reasoning. To improve the precision of fault diagnosis and reasoning, airborne PHM includes a subsystem manager (sensors allocation and fault feature extraction), regional manager (hydraulic system, fuel system etc. abnormal/diagnosis/prediction reasoning), and aircraft manager (aircraft abnormal/diagnosis/prediction reasoning). In the design of airborne PHM, the

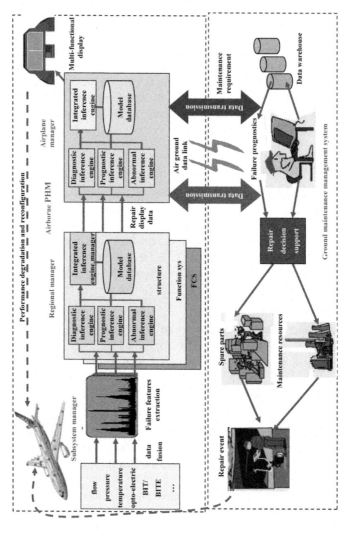

FIGURE 4.43   Aircraft PHM system structure.

multiple sensors network should be allocated according to the failure mode and effect analysis. The sensors extract the failure features and input them to the regional manager for reasoning. The manager adopts the advanced reasoning and data fusion techniques to monitor the system health condition, performance degradation and failure states, repair requirements, and residual useful life evaluation. The failure detection then adopts the robust failure feature extraction methodology. Failure diagnosis adopts the intelligent fault diagnosis algorithm to determine the fault location through hierarchical clustering and cross-correction with a high-fault diagnosis precision and low alarm rate. The regional manager gathers all the information from subsystems, carries out the abnormal/diagnosis/prediction reasoning, and transmits the result to the aircraft manager. The aircraft manager is embedded in the flight management computer, which carries out the top-level PHM integration through cross-correction and data fusion and determines the failure and reports the critical fault to the pilot. Under this circumstance, PHM carries out the reconfiguration or mission degradation to effectively prevent the spread of the failure. Meanwhile, the fault information and residual useful life of key components will be transmitted to the ground maintenance system so that the ground maintenance system can adjust the spare components and repair equipment to realize the rapid maintenance. The ground maintenance management system receives the information from the air ground data link and manages the repair strategy in advance based on the maintenance data warehouse. In certain aircraft systems, PHM can predict the health states with prior knowledge and design model and give the residual useful life. The ground management system determines whether to replace the fault component or impounded fault component and undertake rapid maintenance through expert knowledge database.

The advantages of PHM are to connect the airborne health states to logistic support system, provide the real-time system information to the pilot or repair personnel, and generate the logistic support and rapid maintenance.

PHM includes many disciplines such as fault physics, fuzzy mathematics, pattern identification, and artificial intelligence. Its key technologies consist of advanced sensors, robust fault detection, intelligent fault diagnosis, data fusion, artificial intelligence, data mining, and fault prediction. The following sections will introduce the key techniques of PHM.

4.5.3 PHM Information Acquisition and Fault Feature Extraction

Aircraft PHM information acquisition comes either from the onboard sensors network or from flight data and maintenance data shown in Figure 4.44. The sensors used in aircraft require small size, lightweight, antielectromagnetic interference, and ready connection with a processor. The commonly used sensors include fiber optic sensors, wireless sensors, virtual sensors, intelligent

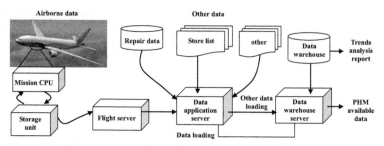

FIGURE 4.44   Data resources of an aircraft PHM system.

sensors, and piezoelectric sensors [44,45]. These sensors are sensitive to failure and antielectromagnetic inference. In addition, aircraft can also load flight and maintenance data into a PHM database.

To pick up the health state and fault features of aircraft system, the essential sensors should be installed in aircraft. PHM gets the information from these sensors and extracts the fault features. Because there are a number of different interferences around these sensors, it is necessary to identify a method to extract the fault features effectively and eliminate the disturbance and highlight the failure features.

The common methods of fault feature extraction used in PHM include wavelet analysis [46], cepstrum envelope [47], empirical mode decomposition method (EMD) [48], and the chaos-based weak signal extraction method [49] shown in Figure 4.45. Herein, wavelet analysis method carries out the multiresolution orthogonal transformation of detected signal, calculates the detailed decomposition in the frequency domain by adopting the multiple filters, and highlights the fault features [50]. The cepstrum envelope method carries out the frequency analysis of detected signals, obtains the envelope with Hilbert transform, and eliminates the disturbance influence [51]. EMD decomposes the detected signals to limit the number of inherent modes function, carries out the Hilbert transform of the selected modes, and determines the fault features. EMD is also known as Hilbert-Huang transform [52]. Chaos-based weak signal extraction method carries out the cross-correlation to eliminate the noise influence, determine whether the detected signal with the periodic signal through the chaotic system variance of phase locus, and improve the signal to noise immunity [53].

4.5.4 PHM Hierarchical Intelligent Fault Diagnosis Algorithm

It is not uncommon to make a wrong judgment if only one signal or one criterion is used in fault diagnosis. As opposed to the traditional BIT fault diagnosis, PHM adopts the hierarchical intelligent reasoning mechanism and cross-correction technology to ensure the high-accuracy fault diagnosis. The specific approach is to carry out a level of reasoning, in which each level

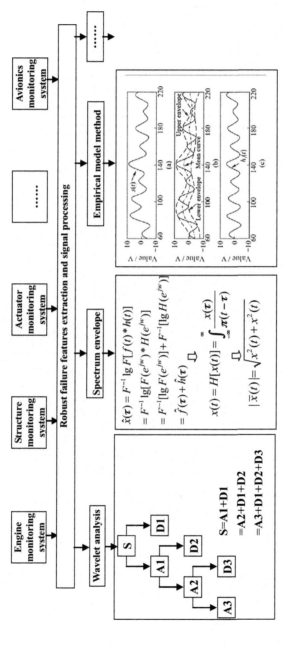

FIGURE 4.45 Robust failure feature extraction from the detected signal.

adopts diagnosis method of their own and cross-correction between levels, to ensure the accuracy of fault diagnosis, and to reduce the alarm rate.

4.5.4.1 Hierarchical Intelligent Reasoning

According to the hierarchical PHM structure, PHM can be divided into a subsystems level, regional manager level, and aircraft manager level, as shown in Figure 4.46. Subsystems include BIT data, parameter acquisition, subsystem fault diagnosis, and subsystem management. The regional manager can perform cross-correction and cross-validation, except the fault diagnosis and prognostics for subsystems. Each level performs its own fault diagnosis function, cross-corrects the detected results with neighbor levels, and obtains the final diagnosis results through cyclic iteration.

Here MS means multiple systems, and A/V means air vehicle.

Aircraft-detected information consists of an on-off signal, continuous signal, and performance degradation, so PHM designs three types of reasoning machine which are described below.

- Abnormal inference machine: For the 0−1 signal or slow varying signal, PHM compares the detected signal with the threshold to determine whether the detected signal exceeds the threshold. If the signal does not exceed threshold, the system is considered to be normal.
- Fault diagnosis inference machine: For a fast varying signal, PHM designs the intelligent fault diagnosis algorithm to determine the system failure. The final failure result is reported to the pilot, and PHM isolates the fault component to the replaceable unit.

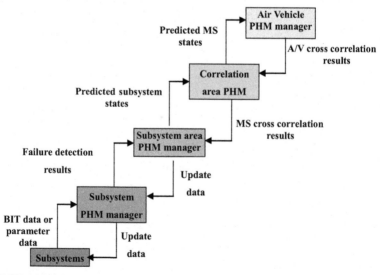

FIGURE 4.46   Hierarchy cross-correction of aircraft PHM.

- Failure prediction inference machine: PHM can estimate the fault time and transmit the maintenance requirement to the ground maintenance center based on the system health state according to the system performance degradation.

4.5.4.2 Enhanced Cross-Correction

To improve the failure reasoning accuracy, PHM adopts the cross-correction methods at different levels to improve the fault diagnosis accuracy [54]. The cross-correction adopts the model-based inference between levels shown in Figure 4.47, in which the component provides the function, sensor detects if the function works and state controls the function operation. Figure 4.47(b) provides the flow diagram of cross-correction of redundant hydraulic system, where solid lines surrounding the section indicate the isolated failure.

In Figure 4.47, EDP means the engine-driven pump fault, P1 means the hydraulic system 1 fault, P2 means the hydraulic system 2 fault, and P3 means the hydraulic system 3 fault.

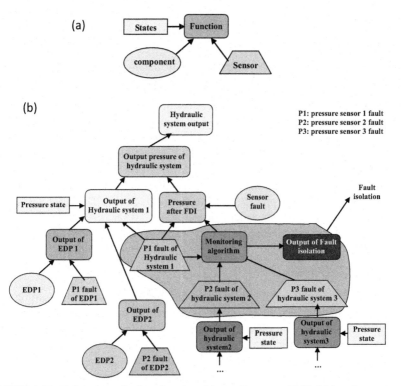

FIGURE 4.47 Failure determination of an aircraft hydraulic system. (a) Model-based reasoning, (b) cross-correction of aircraft hydraulic system.

4.5.4.3 Intelligent Fault Diagnosis Algorithm

The fault diagnosis algorithm of aircraft PHM depends on the artificial intelligent technologies. The common intelligent fault diagnosis methods are listed as follows.

- Fault diagnosis based on the artificial neural network

Under the normal operating conditions, features are collected from the system to train the neural network and obtain the neural network weights. This type of neural network is considered the system model [54]. In real applications, the output of the system is compared with the neural network model and evaluated for the system fault when the error between the system output and output of neural network is greater than the specified threshold, Figure 4.48.

- Fault diagnosis based on support vector machines

Support vector machines (SVM) are the optimal solution to pursue the finite samples based on the minimum structure risk. In essence, SVM is a classification method, which determines the fault and its location according to the information during flight (such as vibration spectrum, waveform, and relevant operation parameters). SVM can obtain good classification results under the conditions of small training samples, as shown in Figure 4.49.

In Figure 4.48, R_i is the failure area i ($i = 1,2,3$) and ω_i is the weight of R_i.

- Fault diagnosis based on multiple sensor fusion

Data fusion is a method that synthesizes multiple sources of information and generates more accurate and more complete estimation judgment than a single information source. Data fusion includes fusion methods based on weighting, parameter evaluation, D-S evidence reasoning, Kalman filter, neural network, and rough set theory [55]. PHM adopts the integrated data fusion method shown in Figure 4.50.

This structure can fuse the initial sensor data and feature data at the same time. In the data fusion process, it is important to extract the useful data from the sensors and improve the calculation efficiency. Then, the independent fault classification algorithm is applied to deal with the features and realize the fault isolation.

- Fuzzy logic inference

Fuzzy logic inference is the method that fuses system input based on the degree of membership function and generate the output [56]. Figure 4.51 shows the flow diagram of fuzzy logic inference. The first step is to take the features and determine the degree to which they belong to each of the appropriate fuzzy sets via membership functions. The second step is to calculate the results with summation or maximizing reasoning algorithm. The third step is to carry out the defuzzification to get the precise output. The last

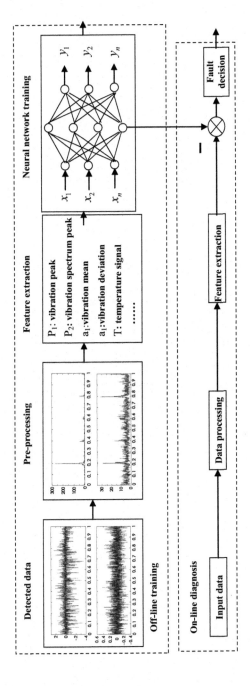

FIGURE 4.48 Fault diagnosis based on neural network.

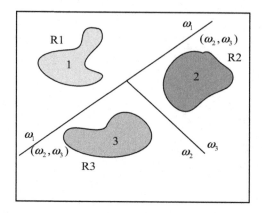

FIGURE 4.49 Fault classification based on a support vector machine.

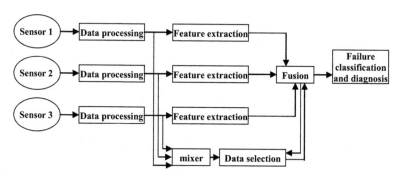

FIGURE 4.50 Data fusion fault diagnosis based on multiple sensors.

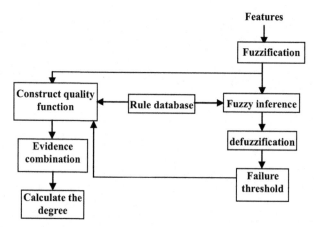

FIGURE 4.51 Fuzzy logic diagnosis.

TABLE 4.5 Failure Mode and Effect Analysis

Failure Mode	Failure Reason	Failure Characteristic	Severe Degree
Lack of inlet pressure	Supercharging device breakdown, liquid level of oil tank is too low, oil viscosity is too high, inlet pipe is too small	Intermittent violent vibration, abnormal noise, fluctuation of outlet pressure is noticeable	III
Wear of port plate	Oil pollution, insufficient lubrication	Volumetric efficiency decrease, increase in oil leakage abnormal vibration	III
Fault of roller bearing	Roller fatigue wear, components of bearing wear and strain	Slight abnormal vibration	IV
Off-centered swashplate	Assembling error, serious wear	Slight abnormal vibration	IV
Increased clearance of piston/shoe	Wear of matching surface between sliding shoe and piston, clearance increment	Slight vibration, no change in performance	IV

step is to obtain the failure threshold for fault diagnosis. Fuzzy logic inference can deal with different types of uncertainty.

- Fault diagnosis based on the hierarchy clustering of hydraulic pump

Let us take hydraulic pump as an example. According to the operational principle, the main causes of failure of hydraulic pump include wear, fatigue, and cavitation. After extensive investigation, five types of common failures are found, which are listed in Table 4.5. Their fault mechanisms are analyzed in detail as follows.

4.5.4.3.1 Lack of Inlet Pressure

Under normal conditions, hydraulic oil in an oil tank dissolves a certain amount of air. When the inlet pressure becomes lower than ambient pressure, air dissolved in oil will come out and form small bubbles that will cause cavitation. In addition, the oil viscosity will decrease rapidly. This phenomenon will affect the normal operation of an airplane, such as abnormal movement of landing gear. When the inlet pressure is low, the characteristic behavior listed below will occur

- Vibration: Intermittent violent vibration will appear during operation, vibration amplitude in the axial and radial directions will increase noticeably.

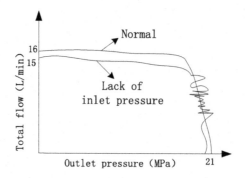

FIGURE 4.52   Total flow outlet pressure curve under low inlet pressure.

- Flow: Under large-flow rate, the sum of discharge flow and pump's oil-return flow is much smaller than at a normal level. Figure 4.52 illustrates this difference.
- Pressure: Fluctuation amplitude of discharge pressure will increase. Under the large-flow rate, amplitude and period of pressure fluctuation are irregular. Under the small-flow rate, fluctuation of discharge pressure is regular. Figure 4.53 shows a partial amplified curve of discharge pressure under small-flow rate.

4.5.4.3.2 Wear of Port Plate

Friction pairing between the port plate and the cylinder barrel is the most important function in a piston pump. Oil film established between a port plate and cylinder is unstable under high pressure and high speed, and dry friction will appear if the oil film is broken. The generated contamination particles resulting from dry friction aggravate adhesive wear and abrasive wear.

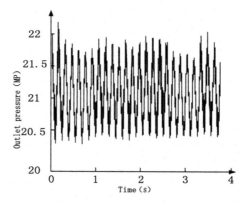

FIGURE 4.53   Partial amplified curve of discharge pressure under small-flow rate.

TABLE 4.6 Relationship of Port Plate Wear and Pump's Oil-Return Flow

Maximum Wear Amount of Port Plate (mm)	Maximum Oil-return Flow of Pump (L/min)
0.006	1.5
0.01	2.6
0.018	3.5
0.026	4.2

A number of experiments have shown that surface morphology of worn port plate is uniform circular scratch. The resulting pump oil-return flow increases noticeably, which is shown in Table 4.6.

4.5.4.3.3 Wear of Roller Bearing

Cylinder support bearing consists of outer ring, roller, and cage. If there is any scratch on the surface of roller, a slight vibration will be produced when the roller bumps against the outer or inner ring during its rotation and revolution. This slight vibration will propagate to the pump housing.

Significant strong environmental noise and strong vibration of a hydraulic system will easily mask the minute vibration produced by the bearing. Research shows that the traditional spectrum analysis method cannot extract roller bearing fault information, whereas the cepstrum envelope method can realize the exact diagnosis of bearing fault.

4.5.4.3.4 Off-Centered Swashplate

Off-centeredness of a swashplate can occur from an assembly error or component wear. When the pump is working, the piston will produce force to the swashplate through the chuck. If the swashplate clearance increases, then the swashplate will move to that side. After a certain offset, the ability of automatic centering of piston components will make the swashplate return to the center. These two effects will eventually lead to back and forth oscillatory movement of the swashplate, and the vibration in radial direction can become abnormal. This type of fault is typical of gradual failure of a pump.

4.5.4.3.5 Increased Clearance of Piston/Shoe

Pistons reciprocate in piston bores with cylinder's rotation. During every cycle, the piston ball-end bumps against sliding shoes twice, and with time, the clearance of piston/shoe increases gradually. When the clearance exceeds a safe level, the piston may disengage from sliding shoe suddenly and destroy the whole hydraulic system. This kind of fault is another type of gradual failure of pump.

Multisensor data fusion technology is usually applied to diagnose a hydraulic system. Information collected by sensors is commonly affected by harsh environments, which make it difficult to realize exact fault feature extraction. For example, when an off-center of a swashplate or clearance of a piston/shoe increases, vibration of a pump's housing will become abnormal but it will be very small compared with a hydraulic system's vibration. Figures 4.54 and 4.55 show a frequency spectrum of vibration in the axial and radial directions when these two faults occur. It is easy to see that the two spectra are similar and that there are no apparent amplitude differences between the two faults. The fault features are coupled, which renders application of traditional diagnosis method to realize accurate fault locating either difficult or impossible.

Fault diagnosis of five common failures described here all depend on pressure, which causes vibration, and also produces substantial noise. Because the signal to noise ratio (SNR) of a sensor's signal is low, useful information is masked, the fault feature is vague, and the accuracy of diagnosis is significantly reduced. Therefore, improving the SNR of signals and eliminating coupling of different faults becomes a key issue in multifault diagnosis.

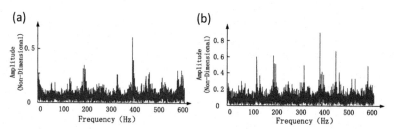

FIGURE 4.54 Frequency spectrum of vibration in the axial direction. (a) Increased clearance of a piston/shoe, (b) off-centered swashplate.

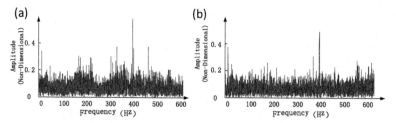

FIGURE 4.55 Frequency spectrum of vibration in the radial direction. (a) Increased clearance of piston/shoe, (b) off-centered swashplate.

4.5.4.3.6 Hierarchical Clustering Diagnosis Algorithm

We need to acquire sufficient fault data through a substantial number of sensors in order to realize accurate and efficient diagnoses. There are various possible failure modes associated with aircraft hydraulic pumps and coupling between them when in failure modes is strong. When multiple faults occur simultaneously, one type of acquired data covers multiple fault features and one type of failure needs multiple measurement information to isolate and diagnose. So, careful selection of sensors and using the fewest possible number of sensors to get the most effective information is the foundation of multifault diagnosis. The optimization of sensor layout is obtained by analyzing characteristics of each fault, t, and is presented in Figure 4.56.

As shown in Figure 4.56, all five types of faults can produce abnormal vibration. Flow and pressure will change if insufficient inlet pressure drops significantly while the pump's oil-return flow will increase noticeably when the port plate wears. We can realize accurate diagnosis of low inlet pressure and wear of a port plate according to these specific characteristics.

Discharge pressure sensors are commonly installed on aircraft. If the total flow decreases under large-flow rate and the fluctuation amplitude of discharge pressure increases under small-flow rate, it can be identified as low inlet pressure. If the pump's oil-return flow increases noticeably, it can be confirmed as wear of the port plate.

When other three types of fault occur, external behavior of pump, such as pressure and flow, does not exhibit obvious noticeable changes. The unique phenomenon in this case is slight vibration of pump housing. Mechanical connection, loosening of connectors, friction pair wear, and bearing damage may all produce pump vibration. Thus vibration contains a wealth of fault information and is the most important signal for diagnosis. As a result, axial and radial vibration sensors are selected to capture this information.

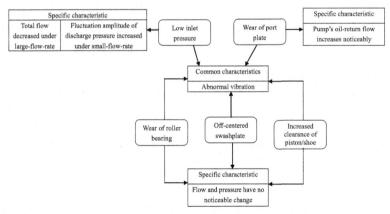

FIGURE 4.56   Characteristics of five failure modes.

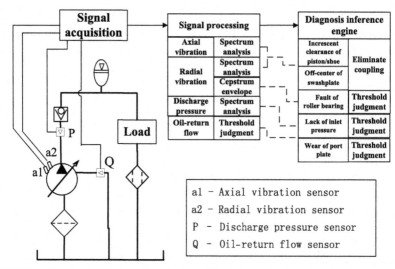

FIGURE 4.57   Optimized sensor layout of a multifault diagnosis platform.

In summary, four types of sensors, including discharge pressure sensor, pump's oil-return flow sensor, and axial and radial vibration sensors, are selected to realize multifault diagnosis. Sensor layout is shown in Figure 4.57.

All five types of faults shown in Table 4.5 can affect pump's performance. When they occur simultaneously, it is necessary to determine an order of diagnosis according to certain principles. A hierarchy of multifault diagnosis is proposed as shown in Figure 4.57. The principle of hierarchy is the fault that causes greatest impact is diagnosed first and the fault with a simpler diagnosis method is diagnosed second.

The first diagnosis level, Figure 4.58, is focused on port plate wear fault and low inlet pressure fault, each of which may cause serious impact on a pump. Port plate wear fault is diagnosed by detecting whether the value of oil-return flow exceeds the threshold. The low inlet pressure fault is determined by examining the amplitude of discharge pressure spectrum. The natural frequency of swashplate is typically less than 20 Hz, according to manufacturer's data. The peak value of discharge pressure spectrum of normal pump appears at 0 Hz. Thus, examine the amplitude of discharge pressure spectrum between 1 and 20 Hz. If the amplitude is much higher than the mean value at other points, then we can conclude that low inlet pressure fault has occurred.

The second diagnosis level is focused on diagnosing the roller bearing fault. In this case, the cepstrum envelope method can be effectively applied.

The third level is focused on increased clearance of piston/shoe and off-centered swashplate faults. Since the characteristics of these two types of fault is not noticeable, it is necessary to find a method to highlight the fault features. The analysis provided previously points to the fact that improving the

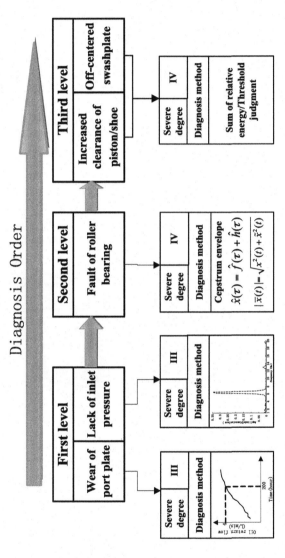

FIGURE 4.58 Hierarchy of multifault diagnosis.

SNR of signals is the key problem of multifault diagnosis. Fault feature extraction based on the sum of relative power can improve SNR and effectively eliminate coupling.

Increased clearance of piston/shoe and off-centeredness swashplate are both gradual failure modes of pump. They have similar fault features and diagnosis methods, and in this case clustering diagnosis should be used. The experiments have shown the following characteristics of vibration spectrum:

- Increased clearance of piston/shoe: Both in axial and radial spectra, the peak value at each harmonic point decreases and the mean value between harmonic points increases. Some peak values disappear because of the fault signal.

- Off-centered swashplate: In the radial spectrum, the peak value at each harmonic point decreases and the mean value between harmonic points increases, whereas the axial spectrum is similar to a normal pump.

For example, the basic axial frequency of a pump is 66.7 Hz; when increased clearance of piston/shoe occurs, axial and radial vibrational spectra are recorded as shown in Figure 4.59. Values of the first to eighth harmonics and the mean value between harmonics are marked in the figure. Figure 4.60 shows the axial and radial vibration spectrums of normal pump.

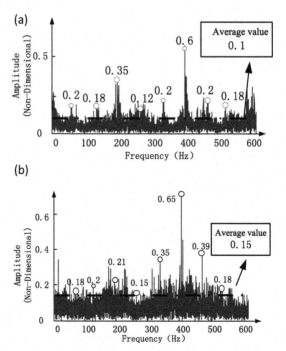

FIGURE 4.59   Increased clearance of piston/shoe. (a) Frequency spectrum of axial vibration, (b) frequency spectrum of radial vibration.

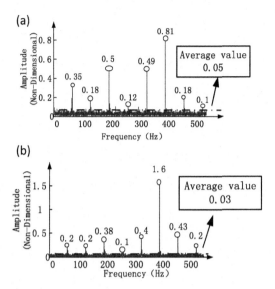

FIGURE 4.60 Normal pump. (a) Frequency spectrum of axial vibration, (b) frequency spectrum of radial vibration.

Comparison of Figures 4.58 and 4.59 indicates that peak values at the second and fourth harmonics are similar, and the value of eighth harmonic of axial vibration increases. This means that although most harmonic values follow the previously mentioned features, there is still randomness. Moreover, although most harmonic values decrease, changes are not apparent. The same conclusion can be made for off-centered swashplate faults. To improve SNR, diagnosis based on the sum of relative power should be adopted shown in Figure 4.61. The steps are as follows.

Step 1: Calculate the power spectral density of axial and radial vibration signals.

Since power spectral density $|X(f)|^2$ is the square of amplitude spectrum $|X(f)|$, high peaks in the spectrum become bigger by the square operation, whereas small value points become smaller. Thereby the purpose of improving SNR is achieved.

In the later text, the power spectral density of axial vibration is designated as $|X_a(f)|^2$, and the power spectral density of radial vibration is designated as $|X_r(f)|^2$.

Step 2: Calculate the sum of relative power and extract fault feature.

In the axial vibration spectrum, choose the first eight harmonics and calculate the average power of each small rang, with every harmonic as the center and ± 10 Hz as bandwidth. Suppose that $\left|\overline{X}_a(\Delta i f_0)\right|^2$ $(i = 1, 2, ..., 8)$

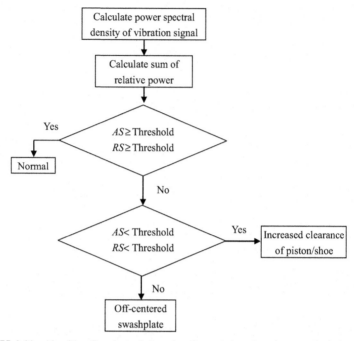

FIGURE 4.61 Algorithm flowchart of clustering diagnosis based on the sum of relative power.

represents the average power of ith harmonic, and f_0 is the basic axial frequency of pump, which is defined as

$$f_0 = n/60 \tag{4.21}$$

where n is the rotational speed of pump with unit rpm.

Then calculate the maximum power value at each harmonic of ± 10 Hz. Suppose that $|X_{a\,\max}(if_0)|^2$ ($i = 1,2,\ldots,8$) represents the maximum power of the ith harmonic point.

Finally, define axial sum of relative power as:

$$AS = \sum_{i=1}^{8} \frac{|X_{a\,\max}(if_0)|^2}{|\overline{X}_a(\Delta if_0)|^2} \tag{4.22}$$

This is the extraction process of axial vibration. Similarly, we can define radial sum of relative power as:

$$RS = \sum_{i=1}^{8} \frac{|X_{r\,\max}(if_0)|^2}{|\overline{X}_r(\Delta if_0)|^2} \tag{4.23}$$

Because most harmonic peak values decrease and the average value between harmonics increases, the relative power of most harmonics would decrease. Thus, by calculating the sum of relative power of harmonics, eliminates the randomness and the sum of relative power AS and RS would always decrease. This process improves SNR further and eliminates random disturbance.

Step 3: Clustering diagnosis based on the sum of relative power (Figure 4.61).

First, analyze the sum of relative power of a normal pump and get an appropriate threshold for diagnosis; then follow diagnosis principles as specified below:

- If $AS \geq Threshold$ and $RS \geq Threshold$, then the pump is normal;
- If $AS < Threshold$ and $RS < Threshold$, then increased clearance of piston/shoe occurred
- If $AS \geq Threshold$ and $RS < Threshold$, then an off-centered swashplate occurred.

Clustering diagnosis algorithm eliminates fault coupling and improves SNR by a simple calculation process that is suitable for real-time diagnosis.

The following section provides verification of the effectiveness and robustness of the hierarchy clustering diagnosis using a multifault diagnosis experiment.

Let us consider an aircraft pump with the following parameters:

- Rated working pressure: 21 MPa
- Rated inlet pressure: 0.4 MPa
- Rated rotational speed: 4000 rpm
- Rated maximum oil-return flow: 2.5 L/min
- Number of pistons: 9

To verify the presented diagnosis method, the faults are deliberately introduced into the system.

4.5.4.4 The First-Level Fault Diagnosis

Because wear of the port plate will result in increased oil-return flow, the pump will be determined as wear of the port plate if the oil-return flow is greater than 2.5 L/min. In this case a "damaged" port plate was installed in the pump. "Damage" consisted of a scratch on the surface of one roller that was made by high-speed milling machine. The scratch had dimensions of $2.48 \times 0.68 \times 0.19$ mm. To diagnose the port plate wear fault, the pump worked under rated conditions, with the maximum oil-return flow 4.2 L/min, which is 1.68 times the rated value. Thus, the wear of port plate can be diagnosed with oil-return flow.

If the inlet pressure is lower than 0.15 MPa, cavitation will occur in the pump. Figure 4.62 shows the spectrum in which the solid and dashed lines represent a normal pump and a fault pump, respectively. The peak value point at 6.7 Hz indicates the fault pump, and fault diagnosis indicates low inlet pressure.

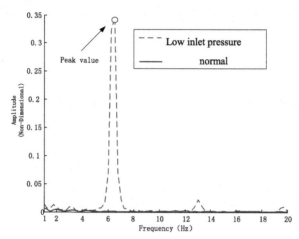

FIGURE 4.62   Diagnosis of low inlet pressure [57].

4.5.4.5 The Second-Level Fault Diagnosis

In the second level, wear of bearing is diagnosed with a cepstrum envelope. According to the bearing fault frequency formula [57], the characteristic frequency of fault bearing is 597 Hz. Figure 4.63 shows the cepstrum envelope curve with is a peak value point at 598 Hz, so diagnosis indicates the faulty roller bearing.

4.5.4.6 The Third-Level Fault Diagnosis

Finally, the increased piston/shoe clearance is diagnosed by applying a clustering diagnosis algorithm based on the sum of relative power. The clearance was manually widened across 12 different settings as indicated in Table 4.7.

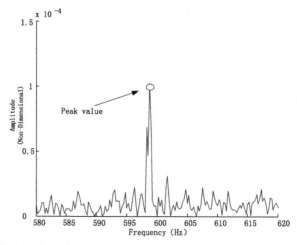

FIGURE 4.63   Diagnosis of roller bearing fault.

TABLE 4.7 Fault Set for Increased Piston/Shoe Clearance

Clearance (mm) Severity / Piston No.	No. 1	No. 2	No. 3	No. 4	No. 5	No. 6	No. 7	No. 8	No. 9
Normal	0.05	0.04	0.04	0.05	0.07	0.04	0.04	0.04	0.04
Level 1	0.06	0.04	0.06	0.06	0.07	0.04	0.04	0.05	0.06
Level 2	0.08	0.06	0.08	0.07	0.09	0.06	0.06	0.08	0.09
Level 3	0.08	0.06	0.08	0.09	0.10	0.08	0.07	0.09	0.09
Level 4	0.12	0.10	0.11	0.12	0.10	0.08	0.09	0.10	0.10
Level 5	0.16	0.12	0.12	0.12	0.13	0.12	0.10	0.13	0.13
Level 6	0.18	0.16	0.12	0.13	0.14	0.13	0.12	0.14	0.14
Level 7	0.18	0.16	0.15	0.16	0.16	0.15	0.13	0.14	0.16
Level 8	0.19	0.18	0.16	0.17	0.18	0.16	0.15	0.16	0.18
Level 9	0.20	0.19	0.18	0.18	0.19	0.19	0.15	0.18	0.18
Level 10	0.22	0.19	0.20	0.19	0.20	0.20	0.17	0.19	0.20
Level 11	0.23	0.20	0.21	0.20	0.20	0.21	0.19	0.19	0.20
Level 12	0.23	0.20	0.23	0.20	0.20	0.23	0.19	0.20	0.20

The rated speed of pump is 4000 rpm; therefore, its basic axial frequency f_0 is 66.7 Hz [57], and the second to eighth harmonic are at 133.3, 199.9, 266.7, 333.3, 399.9, 466.6, and 533.3 Hz. Under large-flow rate and zero-flow rate, the axial and radial sum of relative power are determined from (4.22) and (4.23); the results are listed in Table 4.8. According to values of AS and RS for a normal pump, the diagnosis threshold was set at 100.

The reason for this phenomenon is that the vibration produced by pulsating flow is minimal under zero-flow rate, which makes the impact of fault signal relatively large. However, under large-flow rate, the vibration of pulsating flow is very large, and fault features have a relatively small impact.

When multiple faults occur simultaneously in a hydraulic pump, the hierarchy clustering multifault diagnosis algorithm can significantly improve SNR and eliminate fault coupling. It is very hard to diagnose simultaneous multiple faults by using traditional diagnosis methods.

4.5.5 PHM Fault Prediction

The most significant difference between a PHM and traditional fault diagnosis is the prognostics function. PHM can use the prior fault knowledge and the current state to predict the variation trend of parameters or performance and give the residual life quickly and accurately. With the effective fault prediction, PHM can guide the repair decision effectively before fault occurs and stop the fault development. So the fault prediction can prevent the catastrophic failure after mastering the performance degradation. There are three types of fault prediction methods as which are described in the following sections [57].

4.5.5.1 Model-Based Fault Prediction

There are three types of fault prediction methods: failure physics—based prediction and mathematical model—based prediction.

- Failure physics model—based prediction

Mechanical component failure is based on the failure physics such as wear and tear, fatigue, and aging. Different failure physics has different fault development law. For wear and tear failure, the mechanism obeys the tribology theory [58]:

$$r = kp^m q^n \tag{4.24}$$

where r is the wear ratio, k is the wearing coefficient under specified operating conditions, p is the friction surface pressure, q is the relative speed of friction, and m,n are coefficients. The failure prediction is obtained by integration of Eqn (4.24).

Miner theory is applied in the case of fatigue failure to obtain the failure accumulation law. Suppose that the element operates for n_i cycles under the alternating stress S_i, where the material life cycle is N_i ($i = 1,2,\ldots$). If the fatigue

TABLE 4.8 Diagnosis of Increased Piston/Shoe Clearance

Experimental Group	AS on Axial Direction Under Large-Flow Rate	RS on Radial Direction Under Large-Flow Rate	AS on Axial Direction Under Zero-Flow Rate	RS on Radial Direction Under Zero-Flow Rate
Normal 1	145.92	155.00	475.225	325.43
Normal 2	109.62	241.72	572.976	647.37
Fault 1	83.57	57.27 (Correct)	75.94	76.38 (Correct)
Fault 2	75.27	69.32 (Correct)	63.51	85.24 (Correct)
Fault 3	56.78	76.39 (Correct)	79.85	91.23 (Correct)
Fault 4	44.24	52.65 (Correct)	49.52	45.83 (Correct)
Fault 5	42.86	43.08 (Correct)	44.56	39.65 (Correct)
Fault 6	51.21	54.98 (Correct)	38.83	43.52 (Correct)
Fault 7	43.54	61.04 (Correct)	40.44	38.19 (Correct)
Fault 8	47.19	53.79 (Correct)	43.64	58.62 (Correct)
Fault 9	53.76	169.31 (Misjudge)	70.13	36.25 (Correct)
Fault 10	79.26	70.93 (Correct)	36.48	43.56 (Correct)
Fault 11	63.49	113.61 (Misjudge)	56.59	41.83 (Correct)
Fault 12	86.85	94.42 (Correct)	45.56	43.44 (Correct)

failure is random and if it obeys the exponential distribution, then the failure probability under linear cumulative damage can be described as [59]:

$$\prod e^{-\frac{n_i}{N_i}} = e^{-\sum \frac{n_i}{N_i}} \tag{4.25}$$

According to this equation, the law of fatigue damage can be obtained to understand the failure prediction.

Figure 4.64 shows the prediction process of gear fatigue, whereas Figure 4.65 shows the failure prediction curve of gear fatigue damage.

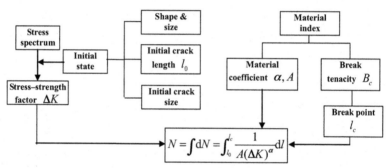

FIGURE 4.64   Prediction process of gear fatigue crack life.

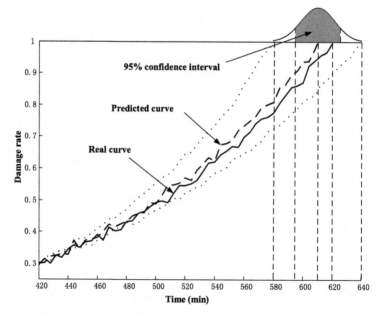

FIGURE 4.65   Failure prediction of gear fatigue damage.

• Mathematical model—based failure prediction

If the mathematical model of the system is known, the stochastic filter theory and particle filter can be used for failure prediction. Suppose the system can be described as

$$\begin{cases} x(t) = f(x(t), u(t), w(t), t) \\ y(t) = g(x(t), u(t), v(t), t) \end{cases} \tag{4.26}$$

where f and g are the system state equation and observation equation of the system respectively, $x(t)$ is the system state variable which is the random variable under certain distribution, $y(t)$ is the system observation value, $u(t)$ is the input variable, and $w(t)$,$v(t)$ are system noise and observation noise, respectively.

The random filter method is to obtain the optimal estimation of system state variable $x(t)$ under certain filtering criterion at time $t = t + \tau$ if the observation sequence is known as $\{y(t), t \in (t_0, t)\}$ from the initial time t_0 to the current time t [59]. If $\tau > 0$, the problem is transformed to the prediction problem. Because the fault evolution can be expressed by state variables, prediction of the state variables can predict the system failure and residual useful life under certain conditions [60].

Particle filter is based on Monte Carlo simulation and recursive bias estimation. Compared with the random filter algorithm, the particle filter is not subject to linear model and Gauss hypothesis constraints, which is theoretically suitable to any system filter problem theoretically [61].

4.5.5.2 Knowledge-Based Fault Prediction

Knowledge-based fault prediction depends on expert knowledge in which its experience are stored in the database. Expert systems can make decisions with knowledge emulating human reasoning and thinking to solve a problem. The goal is to predict a possible failure according to the relationship between the real-time monitoring information and fault severity of the system [62].

Knowledge-based fault prediction depends entirely on the completeness of the knowledge base. The expert system cannot obtain new knowledge from the reasoning operation and thus it does not work for new type of failure. The expert system is more suitable to qualitative reasoning.

4.5.5.3 Data-Driven Fault Prediction

Data-driven fault prediction is the best and most reliable way to predict a failure. However, it is not easy to obtain real data over a long period and the limited data cannot reflect the performance degradation over the life of the system. The common data-driven fault prediction methods are shown in Figure 4.66.

There are three fault prediction methods based on data: support vector machine, wavelet entropy, and progressive failure prediction, Figure 4.66 [64—67]. The multiscale support vector machine is used to integrate the

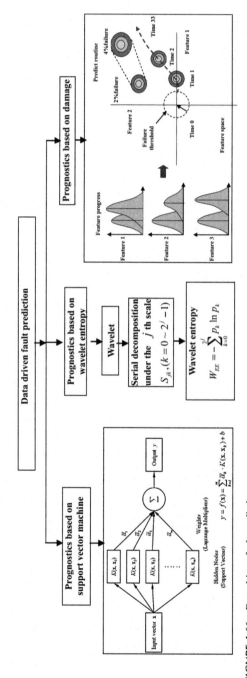

FIGURE 4.66   Data-driven fault prediction.

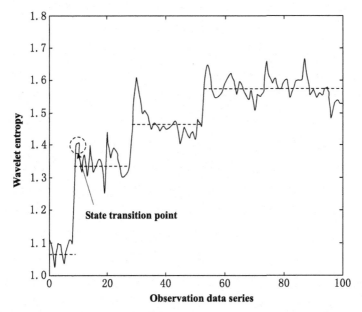

FIGURE 4.67   Prognostics based on wavelet energy entropy.

detected signal [63]. The wavelet entropy is used to obtain the decomposition sequence S_{jk}, ($k = 0 \sim 2^j - 1$) at the jth level and calculate the wavelet entropy W_{EE} to predict the fault after wavelet packet transform. The progressive failure prediction depends on the current features and their change rate of measured element, which can provide the performance degradation and failure information. Since there are some uncertainties in a sensors' output such as boundary condition, material characteristics, damping, and impulse, the sensor output cannot match the response model prediction. It is necessary to eliminate the uncertainties of sensors before using the failure prediction model. Figure 4.67 shows the performance degradation of bearing with wavelet entropy.

4.5.6 Maintenance of PHM

The difference between PHM and traditional maintenance is autonomous logistics support (ALS), which is a new aircraft logistic support system solely based on information and knowledge [68]. In ALS, PHM provides necessary maintenance information to ground maintenance center through monitoring the health status of aircraft hydraulic system. The ground maintenance center can manage the rapid repair through dynamic scheduling repair resources in advance. ALS can decrease repair time, ensure that the aircraft is in good condition, and improve aircraft availability and mission reliability.

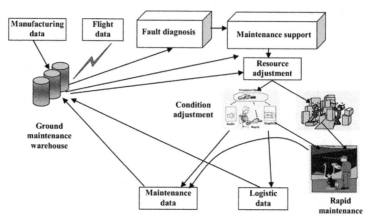

FIGURE 4.68   Ground maintenance management system.

Maintenance supportability is the ability to provide the necessary repair component and resources to recover the system function under certain conditions. ALS not only can provide the accurate failure location, but also can provide the maintenance strategy according to the current health condition.

Normally, the shortest repair time is used to evaluate the maintenance strategy as:

$$T_m = \min f(C, D, P) \qquad (4.27)$$

where T_m is the repair time, C is the repair cost, D is the fault location condition, and P is related to the variable maintenance resources (repair parts and maintenance personnel).

Finally, ALS provides the maintenance suggestion and strategy through interface to repair personnel as shown in Figure 4.68.

In Figure 4.68, core scheduling is a very important issue that can finish the feature extraction, carry out fault diagnosis and prediction reasoning, and adjust the repair resources and personnel.

4.5.7 PHM Evaluation

Aircraft PHM evaluation includes reliability, safety, maintainability, availability, and supportability. There are three parts for PHM: onboard PHM, ground maintenance center, and air-ground data chain. Different parts have different evaluation methods. Figure 4.69 shows the aircraft PHM evaluation methods.

In Figure 4.69, FTA means fault tree analysis, RBD means the reliability block diagram, and FMECA means the failure mode effect and critical analysis. In PHM evaluation, the reliability, safety, maintainability, failure coverage rate, and alarm rate should be comprehensively considered. Considering the

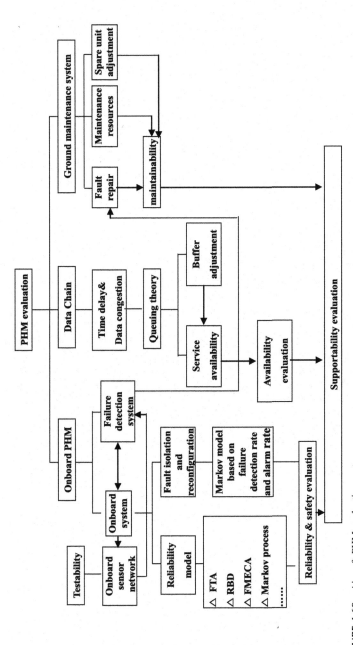

FIGURE 4.69   Aircraft PHM evaluation system.

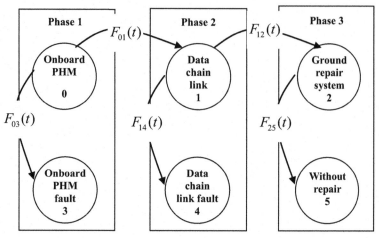

FIGURE 4.70   Markov state transfer diagram of aircraft.

connection among onboard PHM, data chain, and the ground maintenance center, the state transfer diagram of aircraft PHM is shown in Figure 4.70. Its availability can be calculated by using Markovian approach [70].

REFERENCES

[1]   Flightpath 2050 Europe's Vision for Aviation, European Commission. http://ec.europa.eu/transport/modes/air/doc/flightpath2050.pdf.

[2]   B. Chen, Z. Wang, L. Qiu, Main developmental trend of aircraft hydraulic systems, Acta Aeronaut. Astronaut. Sin. 19 (7) (1998) 1–6 (in Chinese).

[3]   J.N. Demarchi, J. Ohlson, Lightweight hydraulic system development and flight test, SAE Tech. Pap. Ser. 801189 (1980).

[4]   H.E. Merritt, Hydraulic Control System, John Wiley, New York, 1976.

[5]   Z. Wang, B. Chen, L. Qiu, Development of aircraft hydraulic system, Hydraul. Pneum. Seal 1 (2000) 14–18 (in Chinese).

[6]   J.H. Brahney, Hydraulic pumps: the key to power generation, Aerosp. Eng. (1991).

[7]   D. Yu, Military Aircraft Hydraulic System, Aeronautic Information Institute, HY95010, 1995 (in Chinese).

[8]   Su-27 for DCS World. https://www.digitalcombatsimulator.com/en/products/flanker/.

[9]   V-22 Osprey Hydraulic Systems, Report No. D-2002-114, Department of Defense Office of the Inspector General. http://www.dodig.mil/audit/reports/fy02/02-114.pdf.

[10] 787 No-bleed systems, Saving Fuel and Enhancing Operational Efficiencies, AERO. http://www.boeing.com/commercial/aeromagazine/articles/qtr_4_07/article_02_3.html.

[11] 8000 psi Hydraulic Systems: Experience and Test Results, Aerospace Information Report, SAE Aerospace Air4002.

[12] Wings design, NASA, National Aeronautics and Space Administration. http://www.aeronautics.nasa.gov/pdf/wing_design_k-12.pdf.

[13] N.H.C. Hwang, R.J. Houghtalen, Fundamental of Hydraulics, Hydraulic Engineering Systems, Prentice-Hall Inc., New Jersy, 1996.

[14] X. Piao, Hydraulic system pressure selection of commercial aircraft, Fluid Power Transm. Control 6 (2011) 22−24 (in Chinese).
[15] P. Tovey, Microprocessor control of aerospace hydraulic pumps, SAE Tech. Pap. Ser. 871863.
[16] J.H. Brahney, Will hydraulic systems meet tomorrow's aircraft power equirement? Aerosp. Eng. 11 (1991) 43−50.
[17] J.E. Spencer, Development of Variable Pressure Hydraulic Systems for Military Aircraft Utilizing the "Smart Hydraulic Pump", C474/002, ImechE, 1993.
[18] S. Wang, Aircraft high pressure hydraulic system and its development, Int. Aeronaut. 2 (1988) 2−8 (in Chinese).
[19] Table 4.1 Hydraulic System Weight Reduction and Fluid Operational Pressure.
[20] W. Zhanlin, L. Peizi, Aircraft Hydraulic Transmission and Servo Control, Defense Industry Press, 1979 (in Chinese).
[21] R.V. Hupp, R.K. Haning, Power Efficient Hydraulic System, vol. 1 (Research) ADA203899.
[22] R.V. Hupp, R.K. Haning, Power Efficient Hydraulic System, vol. 1 (Experiment) ADA203900.
[23] Comparison Diagram on Pump Temperature, Heating Power and Consumed Power between Intelligent Pump and Constant Pressure Variable Pump.
[24] Y. Li, The Design of the Aircraft Intelligent Variable Pressure Hydraulic System Controller and Control Strategy (Master thesis), Beihang University, 2014.
[25] Rongjie Kang. Study on novel electro-hydrostatic actuator and its control strategy (PhD thesis), Beihang University, 2010 (in Chinese).
[26] E.T. Raymond, C.C. Chenoweth, Aircraft Flight Control Actuation System Design, Society of Automotive, first ed., Engineering, Inc., Warrendale, PA, 1993.
[27] D. Scholz, Betriebskostenschättzung von Flugzugsystemen als Beitrag zur Entwurfsoptimierung, in: Proceedings of DGLR-Jahrestagung 1995, Bonn, Germany, 1995, pp. 26−29.
[28] W.E. Murray, L.J. Feiner, R.R. Flores, Evaluation of All-electro Secondary Power for Transport Aircraft, NASA Contrator Report 189077, 1992.
[29] H. Yang, R. Lim, W. Liu, Brushless DC motor modeling and simulation, Micro-motor 36 (4) (2003) 8−10 (in Chinese).
[30] T. Ros, M. Budinger, A. Reysset, J.-C. Maré, Modelica Preliminary design Library for electromechanical actuators, in: Proceedings of AST 2013, 4th International Workshop on Aircraft System Technologies, TUHH, Hamburg, Germany, 2013, pp. 23−24.
[31] C. Meyer, Integrated Vehicle Health Management, National Aeronautics and Space Administration Aeronautics Research Mission Directorate Aviation Safety Program, 2009.
[32] G.B. Aaseng, Blueprint for an integrated vehicle health management system, in: Proceedings of the 20th Digital Avionics Systems Conference, Daytona Beach, Florida, USA, October 14−18, 2001, vol. 1, pp. 3.C.1-1−3.C.1-11.
[33] C. Smith, M. Broadie, R. DeHoff, Turbine Engine Fault Detection and Isolation Program, ADA119998, ADA119999, 1982.
[34] L.A. Urban, Gas Path Analysis of Commercial Aircraft Engine, DFVLR-Mtt 82−02, 1982.
[35] J.G. Early, T.R. Shives, J.H. Smith, Failure mechanisms in high performance materials, in: Proceedings of the 39th Meeting of the Mechanical Failures Prevention Group, National Bureau of Standards, 1984, ISBN 978-0-5213093-9-4.
[36] D.H. Lord, D. Gleason, Design and evaluation methodology for built-in-test, IEEE Trans. Reliab. R-0 (3) (1981).
[37] S. Ofsthub, A approach to intelligent integrated diagnostic design tools, Proc. Atotestcon (1991).
[38] W.E. Hammond, W.G. Jones, Vehicle Health Management, George C. Marshall Space Flight Center, Huntsville, AL, 1992. AIAA: 92−1477.

[39]   A. Hess, L. Fila, The joint strike fighter (JSF) PHM concept: potential impact on aging aircraft problems, in: Proceedings of IEEE Aerospace Conference, Big Sky, Montana, USA, 6, 2002, pp. 3021–3026.

[40]   R.K. Nicholson, K.W. Whitfied, Flight Testing of the Boeing 747-400 Central Maintenance Computer System, 1990.

[41]   M. Davidson, J. Stephens, Advanced health management system for the space shuttle main engine, in: 40th AIAA/ASME/SAE/ASEE Joint Propulsion Conference and Exhibit Florida, AIA, 2004, pp. 2004–3912.

[42]   F.A. Zuniga, D.C. Maclise, D.J. Romano, Integrated system health management for exploration systems, in: 1st Space Exploration Conference, Florida, AIAA, 2005, pp. 2005–2586.

[43]   A. Hess, L. Fila, The joint strike fighter (JSF) PHM Concept: Potential impact on aging aircraft problems, paper # 403, in: IEEE Conference, March 2001.

[44]   M. Dowell, C. Sylvester, Turbo machinery prognostics and health management via eddy current sensing: current developments, in: Aerospace Conference, 1999, pp. 1–9.

[45]   P.J. Boltryk, Intelligent sensor—a generic software approach, J. Phys. Conf. Ser. 15 (2005).

[46]   N. Ricker, Wavelet contraction, wavelet expansion and the control of seismic resolution, Geophysics 18 (4) (1953). http://dx.doi.org/10.1190/1.1437927.

[47]   B.P. Bogert, M.J.R. Healy, J.W. Tukey, The frequency analysis of time series for echoes: cepstrum, psuedo-autocovariance, cross-cepstrum and saphe cracking, in: M. Rosenblat (Ed.), Proceedings of the Symposium on Time Series Analysis, Wiley, NY, 1963, pp. 209–243.

[48]   N.E. Huang, Z. Shen, S.R. Long, et al., The empirical mode decomposition and the Hilbert spectrum for nonlinear and non-stationary time series analysis, Proc. R. Soc. Lond. A (1998) 903–995.

[49]   Y. Wesu, Research on Chaos Theory and Method of Weak Signal Detection, Jilin University, 2006 (in Chinese).

[50]   S. Policker, A.B. Geva, A new algorithm for time series predication by temporal fuzzy clustering, in: Proceedings of 15th International Conference on Pattern Recognition, 2000, pp. 732–735.

[51]   V. Hulle, M. Marc, Kernel-based probabilistic topographic map formation, Neural Comput. 10 (7) (1998) 1847–1871.

[52]   G.L. Clark, J.L. Van, et al., Multi-platform airplane health management, in: Aerospace Conference, 2007, pp. 1–13.

[53]   L. Hongmei, Rotor Fault Diagnosis and Research of Dynamic Balance Adjustment Method of Helicopter, Beihang University, 2010 (in Chinese).

[54]   Z. Anhua, Information Fusion Technique in Equipment Fault Diagnosis, Mechanical Science and Technology, B16, July 1997.

[55]   M. Yuelong, Z. Wu, To explore the methods of dynamic Bayesian network inference based on information fusion, Ship Electron. Eng. 30 (3) (2010) 67–84 (in Chinese).

[56]   L.A. Zadeh, Fuzzy Logic and Approximate Reasoning, Springer, 1975.

[57]   J. Du, Dynamic Fault Mechanism and Health Management of Aircraft Hydraulic Pump (Ph.D. degree thesis), Beihang University, 2012.

[58]   S. Zeng, M.G. Pecht, W. Ji., Prognostics and Health Management Development, 26 (5) (2005) 626–632.

[59]   R.M. Christensen, An evaluation of linear cumulative damage (Miner's law) using kinetic crack growth theory, Mech. Time-Depend. Mater. 6 (4) (2002) 363–377.

[60]   D. Chelidze, A Nonlinear Observer for Damage Evolution Tracking, The Pennysyvania State University, 2000.

[61] M. Orchard, B. Wu, G. Vachtsevanos, A particle filter framework for failure prognostics, in: Proceedings of WTC 2005, World Tribology Congress III, USA, ASME, 2005, pp. 1–2.

[62] E. Lembessis, G. Antonopoulos, R.E. King, C. Halatsis, J. Torres, CASSANDRA: an on-line expert system for fault prognosis, computer integrated manufacturing, in: Proceedings of the 5th CIM Europe Conference, 1989, pp. 371–377.

[63] S. Wang, Prognostics and health management technology of commercial aircraft, Acta Aeronaut. Astronaut. Sin. 35 (6) (2014) 1–12.

[64] T. Biagetti, E. Sciubba, Automatic diagnostics and prognostics of energy conversion processes via knowledge based systems, Energy 29 (12) (2004) 2553–2572.

[65] Z. Yang, J. Guo, et al., Multi-scale support vector machine for regression estimation, Adv. Neural Networks 3971 (2006) 1031–1037. Lecture Notes in Computer Science.

[66] C. Chen, Wavelet energy entropy as a new feature extractor for face recognition, in: The 4th International Conference on Imagine and Graphics, 2007, pp. 616–619.

[67] M.J. Romer, J. Ge, A. Liberson, Autonomous impact damage detection and isolation prediction for aerospace structure, in: Aerospace Conference, 2005, pp. 1–9.

[68] E. Crow, Automatic Logistics Supporting Tactical Combat Vehicles in the MAGTF, Penn State University, 2009.

[69] Bochao Huang. Study on key techniques of aircraft intelligent variable pressure hydraulic system (PhD thesis), Beihang University, 2012 (in Chinese).

[70] S. Wang, Reliability engineering, Beijing University of Aeronautics and Astronautics Press, 2000 (in Chinese).

Abbreviations

2H/2E	Two hydraulic power supply and two electrical power supply systems
2H	Two hydraulic power supply system
ACMP	Alternating current motor pump
ACT	Active control technique
ADP	Air turbine-driven pump
ALS	Autonomous logistics support
BIT	Built-in-test
CA	Criticality analysis
CALCE	Center for advanced life cycle engineering
CBIT	Continuous built-in-test
CCA	Common cause analysis
EBHA	Electro-backup hydraulic actuator
EDP	Engine driven pump
EHA	Electro-hydrostatic actuators
EMA	Electromechanical actuator
EMD	Empirical mode decomposition method
EMI	Electromagnetic inference
FBW	Fly-By-Wire
FMEA	Failure mode and effect analysis
FMECA	Failure mode effect and critical analysis
FO	Failure operation
FPVM	Fixed pump displacement and variable motor speed
FS	Failure safe
FT	Fault tree
FTA	Fault tree analysis
HA	Hydraulically powered actuator
HUMS	Health and usage monitoring system
IVHM	Integrated vehicle health management
LVDT	Linear variable differential transformer
MA	Markov analysis
MCS	Minimal cut set
MTBF	Mean time between failure
MTTF	Mean time to failure
MTTR	Mean time to repair
NRV	Non-return valve
PHM	Prognostics and health management
PTU	Power transfer unit
RAT	Rotary air turbine
RBD	Reliability block diagram

SNR	Signal-to-noise ratio
SOV	Shutoff valve
SV	Servo valve
SVM	Support vector machines
VPFM	Variable pump displacement and fixed motor speed
VPVM	Variable pump displacement and variable motor speed

Notation and Symbols

A	The piston area
A_c	Piston area of cylinder
A_e	Effective area of piston
A_h	Piston area
AS	Axial sum of relative power
B_e	The viscous damping coefficient of cylinder
B_m	Total load damping coefficient of motor-pump module
B_{pe}	Viscous damping coefficient of EHA
B_{ph}	Viscous damping coefficient
C	Repair cost
C	Initial laminar correction coefficient
C_{sh}	Total leakage coefficient
C_{el}	Total leakage coefficient
C_{ele}	External leakage coefficient
C_{eli}	Inner leakage coefficient
C_{pl}	Total leakage coefficient
C_{st}	The leakage coefficient of cylinder
D	Fault location condition
D	Probability distribution of system parameters
D_f	The diameter of piston distribution in cylinder barrel
d	Piston diameter
d_z	The diameter of piston
E_e	Bulk modulus of elasticity
E_y	Equivalent volume elastic modulus
F_e	Output force
F_h	Output force of HA
$F_{\max}$	Maximum force of actuator
f_0	Basic axial frequency of pump
f_e	Damping force
f_h	Friction between piston and cylinder
$G_v(S)$	Transfer function of servo system controller
H	Material hardness
I	Rotor moment of inertia
i	The input current to servo valve
i_e	Motor current
i_v	Input of servo valve
i_{sv}	Current in servo valve
J_c	Moment of inertial of swash plate variable mechanism
J_m	Total moment of inertia of motor-pump module
K_1	Leakage coefficient of the pump

K_c	Flow-pressure coefficient
K_e	Total pressure-flow coefficient
K_e	Opposing electromotive force coefficient
K_f	The feedback gain
K_m	Electromagnetic torque constant
K_t	Connection stiffness of surface
K_q	Flow amplification coefficient
K_s	Abrasive wear coefficient
K_{sv}	The servo valve gain
K_u	The gain of electrical amplifier
K_v	Gain of amplifier
k	Wearing coefficient under specified operating conditions
k_q	Flow gain
L	Distance between variable cylinder axis to the center of swash plate
L	Travel distance of piston
L_e	Armature inductance
L'	Sliding distance
M_f	The equivalent mass
m_e	Equivalent inertia
m_d	Equivalent mass
m_{ph}	Piston mass
m_{pe}	Mass of EHA
N	Gear reducer ratio
n	Number of repairs
n	Number actuation cycles
P	Related to the variable maintenance resources
P_1	Volume loss of pump
P_1	Inlet pressure of cylinder
P_2	Outlet pressure of cylinder
P_a	Outlet pressure
P_b	Inlet pressure
P_e	Load pressure
p	Crew pitch
p	Friction surface pressure
p_f	Load pressure
p_h	Load pressure
p_L	Load pressure
p_n	Designed parameters
p_s	Outlet pressure of pump
Q_1	Flow loss of hydraulic pump
Q_e	Load flow rate
Q_{il}	The internal leakage
Q_L	Load flow rate
$|Q_L|$	Flow that load take from the pump
Q_h	Load flow
Q_t	Flow rate of pump
q	Relative speed of friction surface
q_1	Inlet flow rate of cylinder

q_2	Outlet flow rate of cylinder
q_t	Angular displacement
RS	Radial sum of relative power
R_e	Armature resistance
R_i	Failure area
R_T	Components reliability
r	Ratio of wear
S	The piston stroke
s_0	Stress
T	Life of item
T_e	Output torque of motor
T_i	Life of the ith component
T_m	Torque of motor
T_m	Repair time
T_{sv}	The time constant of servo valve
t	Operational time
t_i	Operational time between failures of repairable product
u_e	Control voltage
u_e	Motor control voltage
V	Displacement of pump
V_c	Total volume of variable cylinder
V_e	Total volume of cylinder
V_i	Performance parameter
V_p	Coefficient of pressure disturbance torque
V_S	Volume of pump outlet
V_t	The volume of cylinder
V_{th}	Total volume of HA
$V_{\max}$	Maximum speed of actuator
W	Normal load
X	System parameters set
$\lvert X(f) \rvert$	Amplitude spectrum
$\lvert X(f) \rvert^2$	Power spectral density
$X_{\max}$	Maximum displacement of actuator
X_t	The piston displacement
x_e	Displacement of piston
x_h	Displacement of cylinder
x_h	Displacement of ha
x_t	Surface displacement
x_v	Spool displacement of servo valve in swash plate variable mechanism
Y	System performance
Z	The number of pistons
α_j	Failure mode ratio
β_e	Fluid equivalent modulus of elasticity
β_j	Conditional probability of mission loss
γ	Angle of swash plate
ξ_h	Relative damping coefficient of swash plate variable mechanism
ξ_v	Damping coefficient of servo valve
ς_h	The hydraulic damping coefficient

ω_e	Motor speed
ω_h	Inherent frequency of swash plate variable mechanism
ω_i	Weight of R_i
ω_m	Motor speed
ω_S	Volume lag tuning frequency
ω_v	Characteristic frequency of servo valve
μ	Repairable rate
μ	Mean or expectation of the distribution
μ	Fluid dynamic viscosity
μ_s	Mean value of of stress s
μ_y	Mean value of random variable y
μ_δ	Mean value of strength δ
σ	Standard deviation
σ_s	Standard deviation of stress s
σ_y	Standard deviation of random variable y
σ_δ	Standard deviation of strength δ
σ^2	Variance
η	Characteristic life described by Weibull distribution
η_v	The volumetric efficiency
Γ	Gamma function
δ	Strength
δ	Single slit height
δ_0	Initial slit height
λ	Failure rate of component
λ_V	Failure rate of voting machine
Ω	Performance threshold
Δ	Leakage flow rate
ΔP	Pressure difference between two chamber
Δp_i	Variance of the ith design parameter
Δp_u	Pressure drop of servo valve
ΔV	Wear volume
ΔV_i	Variance of V_i

Index

Supportability, 115—116, 116f, 118, 129f,
    215—217, 246
  maintenance supportability, 246
System
  reconfiguration, 139
  reliability design, 138—145, 168
  safety requirements, 34—36, 59
  stability, 117, 179—180

T

Testability, 115—118, 116f
Thermal design, 128—129, 129f

Thrust reversers, 35, 59, 61—62, 64f,
    66, 213
Tolerance design, 128, 129f
Transfer function, 46—47, 99—101, 160—161,
    164, 197, 208
Trimmable horizontal stabilizer, 8, 12, 28f, 67

W

Water proof design, 129f
Wear reliability, 136—138
Wear-out failure, 120—121
Weibull distribution, 123—124, 126

Printed in the United States
By Bookmasters